任意设计
模式
建筑和体块
通用元素
建造
结构
立面
全球化设计

Any Design
Patterns
Buildings and Volumes
Generic Elements
Construction
Structures
Facades
Global Design

建筑设计过程通常也被认为是艺术创造的"黑箱"解析过程，人们不能从外部直接窥视其内部的状态，唯一分析的方法是从输入与输出两端揣测其结构与参数。随着数字技术的介入，设计"黑箱"过程被逐步细化，并通过计算机程序算法理性呈现，全新的系统方法使得该过程新颖独特，这不仅体现在最终成果的形态，其设计手段和思维方式的变革也在悄然发生。"程序黑箱"解析融入理性数据、视觉审美及多样元素的互动与形变，并借助计算机程序内部运行机制对阶段性结果反复推敲，"程序黑箱"取代"设计黑箱"必将大幅度降低传统手法在众多候选方案之间来回游弋的操作成本。

本书展示 2000 年至 2009 年瑞士联邦理工大学以 Ludger Hovestadt 教授为首的建筑数字技术研究所的科研轨迹，其学科涉猎已大大超越普通"建筑人"的科研范畴。书中大部分案例的完成者均历经欧洲严格的建筑学研习，然而他们保持对貌似客观的绝对真理的戒心，敢于质疑既有的建筑体系，在孜孜于完美的终极答案的同时也始终保持其多义性和开放性。

李飚

L. Biao

2015 年 9 月 29 日

ETH Zürich CAAD Ludger Hovestadt

国家自然科学基金资助项目（项目批准号：51478116，51408123）
"十二五"国家重点图书出版规划项目
城 市 与 建 筑 遗 产 保 护 试 验 研 究

超越网格——
建筑和信息技术
建筑学数字化应用

Beyond the Grid —
Architecture and Information Technology
Applications of a Digital Architectonic

Ludger Hovestadt 编著
李飚　华好　乔传斌 译　李荣 校

东南大学出版社·南京
Southeast University Press · Nanjing

Prof. Dr. Ludger Hovestadt (Ludger Hovestadt 博士、教

Ludger Hovestadt 博士、教授

　　Ludger Hovestadt教授，苏黎世联邦理工大学CAAD实验室主任。
　　作为一位建筑师和计算机科学家，Ludger Hovestadt的跨学科研究强调"技术与人的互动"。对他而言，信息是反思和重塑人类生活、工作和娱乐环境的关键。因此，他关注于建筑学、计算机科学、机械工程、机器人和机器认知等学科的交叉。他最感兴趣的是能够处理复杂系统的设计与建造工具的开发，尤其是生成设计、数字制造、智能建筑。
　　Hovestadt 教授指出我们有必要也有机会将技术发展与建筑工业相结合，而不只是囿于某一个研究领域，尤其是在建筑节能和建筑经济等领域。他的研究项目因而不仅是技术的实验，更多是寻找一个历史转折点，展示技术在工业实践中的实际应用。自从2000年他成为苏黎世联邦理工大学CAAD实验室教授，他和20个小组成员已经进行了超过100个实验和研究，几个快速发展的附属公司已完成了许多不同类型项目。
　　Hovestadt于1960年出生在德国盖尔森基兴，就读于德国亚琛工业大学(RWTH)，慕尼黑建筑学院，然后在奥地利维也纳设计学院（HFG）师从Holzbauer教授。1987年毕业后，作为研究员在德国卡尔斯鲁厄工业大学和教授Fritz Haller、Niklaus Kohler 教授一起工作，并在1994年获得博士学位。1997年至2000年，Hovestadt 在德国凯泽斯劳滕大学担任建筑计算机辅助设计的客座教授。已婚，有四个孩子，一家人在瑞士苏黎世生活。

Prof. Dr. Ludger Hovestadt

Ludger Hovestadt is Professor for Computer Aided Architectural Design (CAAD) at the Swiss Federal Institute of Technology (Eidgenössische Technische Hochschule, ETH) in Zurich.

As an architect and computer scientist, Hovestadt's approach is interdisciplinary: His concern is to 'interface technology with people.' For him, information is the key to rethinking and reshaping the environments in which we as human beings live, work, and play. He therefore takes an approach aimed at integrating fields as diverse as architecture, computer science, mechanical engineering, robotics, and machine cognition. His particular interests lie in the development of design and construction tools capable of managing complex systems with an emphasis on Generative Design, Digital Production, and Building Intelligence.

Hovestadt identifies both a need for and great opportunities in research that focuses on adapting technological developments from spheres of expertise other than the building industry, especially in the important areas of ecology and economy. His research projects are consequently not only technological experiments, but look for an historical anchor and seek to demonstrate concrete applications for current building practice. Since taking up his Chair for CAAD at ETH Zürich in 2000, Hovestadt and his team of 20 researchers have carried out more than one hundred experiments and studies, which have led to a wide variety of projects in several fast-growing spin-off companies.

Born in Gelsenkirchen, Germany, in 1960, Hovestadt studied architecture at Rheinisch-Westfälische Technische Hochschule (RWTH) in Aachen, Germany, and the Hochschule für Gestaltung (HfG) in Vienna, Austria, with Professor Holzbauer. Upon completion of his diploma in 1987, he worked as a scientific researcher with Professor Fritz Haller and Professor Niklaus Kohler at the Technical University in Karlsruhe, Germany, where he received his doctorate in 1994. Between 1997 and 2000, Hovestadt was a visiting professor for CAAD at the department of architecture at the University of Kaiserslautern, Germany. He is married with four children, and lives in Zürich, Switzerland.

译者简介

李　飚　　　　　　　　　　　　　　东南大学建筑学院　教授　博士生导师

东南大学建筑学院博士，师从钟训正教授、院士，苏黎世联邦理工大学博士后，师从Ludger Hovestadt教授。长期从事与瑞士苏黎世联邦理工大学建筑数字技术中瑞科研合作。

研究领域涵盖建筑设计及其理论、计算机编程技术及建筑数字技术、建筑数控建造等。现任东南大学建筑学院建筑设计本、硕、博教研工作，全国建筑数字技术教学工作委员会副主任，东南大学建筑学院建筑数字技术与应用研究所所长，东南大学城市与建筑遗产保护教育部重点实验室主要成员。

华　好　　　　　　　　　　　　　　　　　　苏黎世联邦理工大学　博士

东南大学建筑学院学士、硕士，2009年加入Ludger Hovestadt教授的CAAD（计算机辅助建筑设计/ETH Zurich）实验室深造并获得博士学位。曾在ETH、欧盟ERASMUS项目中讲授基于计算机编程的实验性建筑设计，2014年起在东南大学开展数字建筑教学。

主要研究设计运算（computational design）和数控建造，基于信息技术对建筑设计方法进行重构。曾获CAADRIA青年研究者奖，国家自然科学基金赞助，在eCAADe, CAADFutures, Computer-Aided Design，《建筑学报》等国内外刊物上发表多篇论文。

乔传斌　　　　　　　　　　　　　　　东南大学建筑学院　建筑学硕士

师从东南大学建筑学院李飚教授，建筑运算与应用研究所成员,主要研究方向为参数化建筑设计（Parametric Design），精通VB等计算机程序语言，以及各类软件的建筑实践应用。

任意设计	9
模式	51
建筑和体块	73
通用元素	93
建造	111
结构	151
立面	183
全球化设计	203

Any Design	9
Patterns	51
Buildings and Volumes	73
Generic Elements	93
Construction	111
Structures	151
Facades	183
Global Design	203

目录

前　言

第一章　新的深层结构   1

任意设计

Schuytgraaf（荷兰）：地块的组织规划   10
Heerhugowaard（荷兰）：发展性规划设计   18
Oqyana（阿联酋，迪拜）：把形式交给数据   22
Globus Provisorium（瑞士，苏黎世）：共识引擎   28
Hardturm（瑞士，苏黎世）：数字链   36

第二章：超越网格   45

模式

The Millipede：拐角   52
Semper Rusticizer：追踪   58
Processing：编程而非绘图   62

第三章：不可预估的设计   67

建筑和体块

Grünhof（瑞士，苏黎世）：投影禁令   74
Stadtraum Hauptbahnhof（瑞士，苏黎世）：建筑作为调节器   80
Bishopsgate（英国，伦敦）：城市的谈判   88

通用元素

Replay Column Atlas：现代古迹   94

第四章：结构与形式的分离   105

建造

Olympia Stadiun（中国，北京）：尝试及错误   112
Futuropolis（瑞士，圣加伦）：数字链效应   116
Swissbau Pavilion：生长的建筑   124
Stadsbalkon（荷兰，格罗宁根）：圆柱的舞蹈   130

目录

    Metrostation（意大利，那不勒斯）：稳定性的颠覆   134
    Monte Rosa（瑞士）：数字小屋   138

第五章：过程和平衡   145

结构

    Pavillons：数字建筑结构   152
    Paravents：学习用机器建造   158
    Freie Innendruck Umformung：吹片   162
    Monster Structures：由纸板制成的屋顶   170

第六章：数据和信息   177

立面

    Südpark（瑞士，巴塞尔）：从内部到外部   184
    Alu Scout：参数化的立面   188
    Credit Suisse（瑞士，苏黎世）：缠绕的／装饰的混凝土   192

第七章：当事物开始学会跑的时候   197

全球化设计

    无处不在的电脑   204
    像鸟儿一样飞翔   208
    建筑谷歌   212
    制造建筑的机器人   218
    在施工现场   224
    电子智能   230
    能源过剩   236
    瑞士一切网络化   242

第八章：虚拟现实的应用   249

    人员   259

Table of Contents

Foreword

Text I: A New Deep Structure   4

   Any Design

   Schuytgraaf (NL): The Organization of Plots   10
   Heerhugowaard (NL): Arranging Historical Growth   18
   Oqyana(Dubai,UAE): Giving Form to Data   22
   GLobus Provisorium (Zurich,CH): The Consensus Engine   28
   Hardturm (Zurich, CH): The Digital Chain   36

Text II: Beyond the Grid   47

   Patterns

   The Millipede: Around the Bend   52
   Semper Rusticizer: Chased Out   58
   Processing: Programming, Not Drafting   62

Text III: The Design of the Unforeseeable   69

   Buildings and Volumes

   Grünhof (Zurich,CH): Forbidden Shadows   74
   Stadtravm Hauptbahnhof (Zurich,CH): Architecture as a Regulator   80
   Bishopsgate (London,UK): Cities Negotiating   88

   Generic Elements

   Replay Column Atlas: Modern Antiquity   94

Text IV: The Separation of Structure and Form   107

   Construction

   Olympic Stadium (Beijing, CN): Trial and Error   112
   Futuropolis (St,Gallen,CH): Digital Chain Reaction   116
   Swissbau Pavilion: A Growing Construction   124
   Stadsbalkon (Groningen,NL): The Dance of the Columns   130

Table of Contents

Metrostation（Naples,IT）:Foreshaking Stability — 134
Monte Rosa（CH）:The Digital Chalet — 138

Text Ⅴ :Process and Balance — 147

Structures

Pavillons: Digital Building Construction — 152
Paravents: Learning to Build with Machines — 158
Freie Innendruck Umformung: Blowing Sheets — 162
Monster Structures: A Roof Made of Cardboard — 170

Text Ⅵ: Data and Information — 179

Facades

Südpark（Basel,CH）:From Inside and Outside — 184
Alu Scout: The Parametrical Facade — 188
Credit Suisse（Zurich, CH）:Winding/Wrapping Concrete — 192

Text Ⅶ :As Things Have Learned to Walk — 199

Global Design

Computers Everywhere — 204
Flying Like Birds — 208
Architectural Google — 212
An Automaton for Making Architecture — 218
On Site — 224
Electrical Intelligence — 230
Energy Abundance — 236
All of Switzerland Online — 242

Text Ⅷ :Applied Virtuality — 253

Persons — 259

前　言

2000年9月我在苏黎世联邦理工大学担任CAAD实验室的教授,当时计算机辅助建筑设计还主要是指在计算机中模拟建筑模型。有两个方向很盛行:一是在"虚拟现实"的名号下——简单来说,就是计算机图形学的应用——建筑的外观特征由电脑模拟,二是以"人工智能"为宗旨的在其深层次结构的模拟。作为一名Fritz Haller教授的学生,我在德国卡尔斯鲁厄大学学习Fritz Haller的建筑构造课程,进行了十年的研究并获得博士学位,因此我对建筑物的结构更感兴趣,而不是它们的图形表达。这使我们的工作在与卡尔斯鲁厄的其他小组进行比较时显得不太一样:我们关心的不是计算机在艺术潜力上的发展,而是更关注它在实际建筑问题上的应用;我们的信条不是"虚拟现实",而是"回到现实"。

一直以来我都想在苏黎世联邦理工大学设立并完善CAAD专业。我们的工作重点始终在于探讨建筑学和数字技术之间的特殊关系。所以,一方面我们正在对决定这一关系的基本的结构性因素进行深入研究,另一方面,我们想把我们的成果放到建筑实践中进行实测。建筑学和数字技术彼此间的互动已经远远超出人们最初的设想,因为这两个领域都更具有综合性而不是单一性。我们的工作是不断地提出新的结合方式以及令人耳目一新的多种解决方式。

在过去的八年中,我们的工作大部分采取的是一种探索性的方法,因此我们并没有将工作内容公之于众。而现在,这样的探索试验阶段已经达到了一个饱和点,所以在接下来的几年里,我们将更多地专注于系统性、理论性和方法论的问题,并且将更多地致力于研究结果的发布和交流上。对我们来说,这本书是迄今为止工作经验的集合,也是今后更关注的理论工作的基础。

通过这本书我们想总结一下我们在CAAD实验室第一阶段的工作。在这个阶段中,我们做了大量丰富的实验,本书精选了其中一部分。它们涵盖了研究中诸多领域的广泛可行性应用,这个领域中数字技术迎合了建筑学的要求。希望这本书可以激励读者进一步深化有关建筑的具体想法。还希望通过这本书,能直观地表现当下建筑学正在发生的一些根本性的变化,读者可能会由此产生很多灵感。但本书未涉及具体操作实践的教程。

本书中的文章尽量做到浅显易懂,既没有项目中所用到技术的深奥解释,也没有理论方面大段的详细阐述。我们展示了一种基于现实世界的实验而存在的广义建筑学的可能性,这令我们非常兴奋。实验的理论和应用可能会在项目叙述间插入的文章中说明,但这只是一个暂定的方式。其详细的说明将是未来研究的一个方向。

我们研究组所拥有的资源,在其他同类甚至国际水平的研究所中均不常见,因此在这里被任命为教授是莫大的荣幸。我们正处在对该学科领域基本问题长期深入研究的位置。目前的卓越成果基于前辈们的努力,尤其是CAAD研究组的创始人Gerhard Schmitt教授。我想借此机会感谢所有前辈,当然,还有所有用热情、精力和技能投入,为如此多层面的项目带来硕果的同事们。

Ludger Hovestadt　/　2009年5月于苏黎世

Foreword

When I assumed the professorship for CAAD at the ETH in Zurich in September 2000, computer aided architectural design was mainly a matter of building models in a computer. Two directions were then prevalent: Under the title of 'virtual reality'—in simple terms, applied computer graphics—the surface features of architecture were reproduced in the computer, while the designation 'artificial intelligence' referred to the reproduction of its deep structure. As a student of Fritz Haller—from whose course in building construction at the University of Karlsruhe I received my doctorate and where I carried out ten years of intensive research—I had a particular interest in the structure of buildings, rather than their graphic representation. This sets our work in Karlsruhe apart in comparison to the other groups: We weren't concerned with developing the artistic potential of the computer, but rather with its use in answering real architectonic questions; our credo was not 'virtual reality,' but 'back to reality'.

It has always been with this goal in mind that I set up and have advanced the professorship for CAAD at ETH Zurich. The focus of our work lay—and still lies—in exploring that special relationship between architecture and information technology. So, on the one hand, we are making a deep study of the fundamental structural determinants of this relationship, while on the other hand, we want to put our outcomes to the test in real-world architectural practice. Architecture and information technology have more to do with each other than one might at first suppose. Because both fields have a somewhat more integrative than particular character, our work has constantly given rise to new hybrid forms and solutions that are as surprising as they are versatile.

Over the last eight years we have taken an exploratory approach to our work, and we have therefore intentionally published and communicated little about it. Now that this experimental phase of applied research has reached a certain saturation point, we will, over the next few years, focus more on systematic, theoretical, and methodical questions, and we will invest more of our resources in publication and communication. For us, this book is a collection of the empirical work that we have carried out so far, and also a basis for our future work, which will focus much more on theory.

With this book we want to conclude this first phase in the development of our work at the chair of CAAD, a phase which was characterized by many and varied experiments, of which we present a selection here. They cover an extraordinarily large area of possible applications in a fertile field of research, in which information technology meets architecture. We hope that this publication will inspire our readers to further develop their own specific ideas about architecture. We furthermore hope that it will help to give a realistic impression of how architecture is currently undergoing fundamental change. Our readers will find much to inspire them, but will find no instruction manuals or step-by-step guides.

The articles in this book have been intentionally kept simple; there are no in-depth explanations of the technologies used in the projects, and theoretical positions are not explained in any great detail. However, we will take great delight in demonstrating the possibilities of an expanded idea of architecture based on our real-world experiments. The theoretical aspects of these experiments and opportunities will be outlined in the text inserts between the descriptions of the projects, but only in a tentative manner. To elaborate on this will be a matter for research in the future.

To be appointed to this professorship is a great privilege, since our chair has access to resources that few other institutes of this type have, even at an international level. We are in a position to carry out thorough long-term research into the fundamental questions of our subject area. This extraordinary position has been made possible by my predecessors; and special thanks must go to the founder of the chair for CAAD studies, Gerhard Schmitt. I would like to take this opportunity to thank all of them, and, of course, all of the many dedicated co-workers who, with enthusiasm, energy, and skill have brought to fruition such a large number of multifaceted projects.

Ludger Hovestadt / Zurich, May 2009

第一章

新的深层结构

　　本书的主题是建筑学和数字技术之间的互动。我们的研究领域称为计算机辅助建筑设计，简称 CAAD。这一领域的常见应用是通过计算机生成建筑结构模型。换句话说，建筑学变得计算机化了。通常情况下，三维建筑模型是可以虚拟地允许人行走通过的。CAAD 的另一个应用研究是算法，它可以一定程度上自动生成建筑或城市模型并使之不断完善。苏黎世联邦理工大学的 CAAD 实验室采取完全不同的方法：我们将建筑放回现实。换句话说，我们正在为当代现实中的建筑实践寻求拓展和改进，这几年来已经取得了一系列创新成果。

　　迅速变化的科技使我们可以将现代建筑看成是实验性的。在计算机的帮助下，壮观、大胆、引人注目的形象和外观设计比以往任何时候更容易。从深层结构中解放出来的外观形成了一项有趣的游戏，这个游戏取决于人们最关注哪一部分。虽然已经进行了许多各种形式的实验，但即便是在最现代的建筑中，建筑物的主要结构构件、基础设施及其维护也并没有太大的改变。

　　我们对表面与结构的分离没有什么兴趣，并且我们也不属于形式反映深层结构这个由我们很多建筑学同事所拥护的阵营。我们更感兴趣的是从长远发展来看，这两者之间的趋同。因此，我们正沿循一条能反映瑞士建筑传统中对结构的特殊想法的道路，同时与英美建筑进行对比。我们所走的路是依据"结构"的，在信息技术的帮助下，可由此将"结构"演进为"叙述性的基础设施"。作为对延伸的建筑准则的补充，这可能要开辟一个全新的建筑学平台。

　　当然，我们并不是声称重构建筑！"建筑即是建筑即是建筑即是建筑。"一种更有效地看待上世纪建筑中出现的大量变化的方法就是采用"大尺度缩放"的方式来看待问题。越来越多的技术发展使我们得以将环境中的对象拆解成非常小的部分，反过来说，也可以利用完全不同的要素来构筑建筑。我们所说的这些可以通过回看生物学发展的历史来解释：在 17 世纪，植物和动物的分类是根据一些很宽泛的特征，比如植物叶子的大小和数量等。但随着显微镜的发明，对单个细胞的研究开始了，比如植物就可被分解得相当小来进行研究，从那以后我们便学会了将它们进行重组——即使只是在原理上、以生理图式的形式。50 多年来，我们观察得更加深入，可以把细胞分解到它的信息载体的程度。现在，我们已可以根据 DNA 来分类生命体，而且在一定程度上，可以通过重组 DNA 对有机体进行构建或重建。但现在我们并不能说已经完全达到解码的程度，因为仅仅是观察的深入，并不意味着这扇门就自己打开了。当我们处理最细微层面的时候，就发现我们并不是在与可接触的实体打交道了，道理很简单——就像锁只有它相应的钥匙才能打得开一样。当我们深入到越细微的层面，比如说分子或更细微的层次，我们就越不容易观察到它们之间的相互作用。我们不得不制定虚拟代码来制定秩序。用直觉来感知世界的状态已面临危机。

建筑历史中也发生过类似的事件：在18世纪，像凡尔赛宫这样的宫殿能以如此的规模和速度建立，主要是因为将数百个相似部件组合在一起——在诸多工厂大规模手工生产。这与需要几十年时间建造的哥特大教堂形成对比，教堂的每个构件都需要手工制作。

凡尔赛宫的建成宣告了工业化成为建筑的必要条件。一百年后，伦敦为万国博览会建成了水晶宫。它由更标准化的单个建筑元素建造出来，关键因素是这些元素可以工业化生产。这不仅是一种生产过程，而且成了一种先进的生产方式，虽然这带来了一些新的限制，但也为建筑开辟了新的可能。今天，我们以拉斯维加斯为例，可以看出重点已经转向了类似DNA一样的解码。如今，代码为沟通、规划、融资、建造的发生提供了基础和框架，并且也正是依靠编码才能完成自由形式和多变的表面。

"建筑即是建筑即是建筑即是建筑即是建筑。"但是建筑表达的媒介却发生了变化：从前工业时代的手工生产到工业化生产再到现今的数字时代的信息通讯。

"大尺度缩放"的一个显著特点就是随着元素的增加，元素间联系的数目也在增加：系统变得更加复杂而不清晰。1947年，随着控制论的出现，美国数学家Warren Weaver提出了一条有趣的推理路线，即区别问题的三个层次。第一层是由少数元素及其之间简单的关系构成的问题，这种问题就可以用简单的符号学逻辑来解决。例如，可以通过力学定律解决的问题就属于这种类型。第二层是较大的问题，由数量庞大的元素、关联和变量——至少看上去是——组成，这只能利用统计方法来描述。这类问题诸如预测人口发展、基本元素在实际多个体系统中的行为，甚至可以是对生物系统的确切看法。第三层面是由那些"迷"一样的问题组成，它们既不能用符号化逻辑来解决——它们包含有太多的元素，也不能用统计方法来解决——包含的元素又少了些。建筑设计就属于第三层面的问题，从这里我们可以开始采用有趣的信息技术。

在工业化社会中，限定的逻辑和简单的算法已开始应用到建筑问题中。大规模生产和高度规律化、重复的立面可以被看做是一种简化表达，也可以被看做是处理复杂建筑问题的保障。民主德国的板片建筑就是这种处理方式的典型（由混凝土预制件建成的高层建筑）。

我们相信，通过数字技术的使用，我们已经找到了当代建筑问题解决策略的关键，也同样适用于历史建筑问题。系统的元素可以任意地被描述和链接，而不失去与实体之间系统化映射关系的一致性。在许多领域，我们可以省掉那些刻板的定义和简化的逻辑，如今在算法的帮助下，我们能够在一个包含大量细节和信息的尺度上描绘建筑问题。我们可以不在乎那些线性的控制或抽象的网格，这样我们才能找到建筑问题更好的答案。

A New Deep Structure

This book has as its theme the interplay between architecture and information technology. Our area of study is known as computer aided architectural design, or CAAD for short. The usual application in this field is the modeling of architectural structures by computer. In other words, the architecture gets 'computerized'. Very often, three-dimensional architectural models are created that can, in a virtual sense, be walked through. Another area of applied research in CAAD is algorithms that can take computer models of buildings or cities and 'grow' them, more or less automatically. The chair for CAAD at ETH Zurich takes an altogether different approach: We situate architecture back in reality. In other words, we are looking for changes, extensions, or improvements to contemporary, real-world architectural practice. Over the last few years, this has yielded a number of innovative results.

The fast-changing nature of technologies available allows us to treat contemporary architecture as an experimental discipline. Spectacular, daring, and attention-grabbing shapes and surfaces can, with the help of a computer, be designed more easily than ever before. This freeing-up of the surface from the underlying structure gives rise to an interesting game, which depends on where the primacy of attention is being directed. While many experiments have been carried out in free form, the treatment of the main structural elements of buildings, their infrastructure, and their maintenance has not changed much, even in the most contemporary architecture.

We are not interested in separating form from (or through) structure, and we also do not subscribe to the school of thought—one espoused by many of our architectural colleagues—that form should again reflect the underlying structure. We are much more interested in the long-term convergent development between the two. Thus, we are following a path that reflects the special significance of ideas about construction within the Swiss building tradition, and that contrasts with the Anglo-American idea of architecture. We are following a path whereby 'structures' can, with the help of information technologies, develop into a type of 'narrative infrastructure'. Complementary to the extended architectonic formal canon, this might open up a whole new plateau for architecture.

Of course, we are not claiming to be reinventing architecture! A house is a house is a house is a house. An easier way of approaching the manifold changes in architecture in the last centuries is to adopt a way of looking at things that we call BIG ZOOM. The ever increasing number of technical developments allow us to deconstruct the objects in our environment into ever smaller parts and, conversely, to build our own artifacts from ever more, and from ever more differentiated, components. What we mean by this can be explained by looking at the history of biology: In the 17th century, flora and fauna were classified according to very broad characteristics: size and number of leaves, for example. With the invention of the microscope, the study of individual cells began; plants, for example, could be broken down into considerably smaller elements, and, since then, we have learned to reconstruct them—even if only in principle, in the form of physiological schemata. For more than fifty years we have been able to look even more closely, and can break cells down to the level of their information-carrying elements. It is now possible to chart organisms at the level of their DNA, and, on a very moderate level, to (re)construct them by recombination. It is not yet possible to speak of a decoding in the actual sense of the word—just because we can look more closely doesn't mean that a door is opening all by itself. When we are concerned with the very finest levels of detail, we recognize that we are no longer dealing with physical, graspable objects, things that are simply there—like a lock with its corresponding key, for instance. The finer the level of detail, down to the domain of individual molecules and beyond, the more the effects of their interplay are hidden from us. We will have to devise virtual codes that we can use to generate order. We have arrived at a crisis in our intuitive ability to view the world.

Analogous events can also be found in the history of architecture: In the eighteenth century, a palace like Versailles could only be built on such a scale and at such speed because, in principle, hundreds of similar parts—created using manual mass production techniques, those of the manufactories—could be put together. This can be contrasted with the decades-long construction time for Gothic cathedrals, which was due to the individual manufacture of the building elements.

With Versailles, production became a necessary condition for architecture. One hundred years later, in London, the Crystal Palace was built for the Great Exhibition. It was built out of individual construction elements that were even more standardized. Crucial for this building was the fact that these elements could now be produced industrially. It was not just the production process, but the means of production that became preeminent and, although it created some new constraints, it also opened up completely new possibilities for architecture. Today, we can look at the example of Las Vegas and see that, in architecture, the emphasis has moved towards codes that are comparable to those found in DNA. Today, codes prepare the grounds as well as the framework upon which communication, planning, financing, and construction take place, and it is also the codes which allow for the free forms and the variegated surface finishes.

A house is a house is a house is a house is a house. But the medium in which architecture expresses itself has changed: from preindustrial manufacture through industrial production to the communication processes of today's information age.

A particular feature of BIG ZOOM is that as the granularity of elements increases, so does the number of connections: the system becomes more complex and less clear. In 1947, with the emergence of cybernetics, the American mathematician Warren Weaver came up with an interesting line of reasoning, distinguishing between three classes of problems. Firstly, simple problems that consist of a few elements and connections, which can easily be solved using simple symbolic logic. For example, problems that can be solved by the laws of mechanics belong to this type. Secondly, the big problems, those that consist of an unmanageable number of elements, connections, and variables—or at least appear to do so—and which can only be described using statistical methods.

Problems of this type involve, for example, prognoses about further development of the population, the behavior of elements in physical n-body systems, or even a certain view on biological systems. The third class is composed of the 'middling' problems, those that cannot be tackled using symbolic logic—they contain too many elements—or statistical methods—they contain too few. To this third class belongs the design of buildings, and that's where the fun with information technology starts.

In industrialized societies, architectural problems have been worked on using restrictive logic and simple algorithms. Mass production and highly regular, repetitive facades can be seen as an expression of simplification, and can be viewed as safeguard systems for managing the complexity of architectural problems. They may have found their most extreme expression in the East German Plattenbauten (high-rise flats built from prefabricated concrete).

We are convinced that by using information technology we have found the key to new solution strategies for contemporary—as well as historical—architectural problems. The elements of a system can be described and linked comparatively freely, in an almost unlimited manner, without losing the consistency of the systematical mapping of the artifact. In many areas, we can dispense with rigid definitions and simplified logics, and we are now, with the help of algorithms, able to map architectural problems on scales that allow for a great richness of details and information. We can leave behind linear ways of controlling and context-free grids, and by doing this, we can find better answers to our architectural problems.

Michel Foucault:
Archaeology of Knowledge

' The problem is no longer one of tradition, of tracing a line,
but one of division, of limits;
it is no longer one of lasting foundations,
but one of transformations that serve as new foundations,
the rebuilding of foundations. '

任意设计

Any Design

项　目：**Schuytgraaf（荷兰）**

时　间：2001—2003

参与者：Markus Braach, Oliver Fritz

合作者：Kees Christiaanse Architects and Planners (KCAP) (Rotterdam, NL)

赞助商：Ministerie van Volkshuisvesting, Ruimtelijke Ordening en Milieubeheer (VROM) (NL); Nederlands Architectuurinstituut (NAi)

地块的组织规划 The Organization of Plots

在CAAD小组所执行的工作项目中，Schuytgraaf项目扮演着一个特殊的角色。因为它是对我们的方法所采用的基本原则的第一次大规模测试，这也是第一次在每一个设计环节都得到信息技术支持的建筑和城市规划工程实践。

The Schuytgraaf project plays a special role in the work undertaken under the banner of the CAAD professorship, since it was the first large-scale test of the basic principle underlying our approach. For the first time, an architectural and city planning project that was aided in each stage of the design process by information technology could be incorporated in practice.

在荷兰阿纳姆Schuytgraaf城郊，一个10公顷名为"12号地块"的基地依据城市规划导则进行发展规划。正如名为"Vinex"的国内建设计划所规定，其框架条件一开始就已明确限定，有很多限定因素：建设地块的组合、不同房屋类型的大小、公共开放空间的数量，以及街道和人行步道的布局。规模包括90个地块上的247栋住宅开发。

In the Arnhem suburb of Schuytgraaf, Holland, the ten-hectare 'Field 12' was to be developed in accordance with city planning guidelines. The existing framework conditions, as set out in the city's 'Vinex' domestic building program, were clearly defined from the outset, with many parameters fixed: the mix of building plots, the sizes of the different house types, the amount of open and public space, and the layout of streets and footpaths. Altogether, 247 dwellings on 90 plots were to be developed.

通常这种布局设计需要借助手工设计：复杂而耗时，且是在不断试错的过程中建立起来。为了使事情简单一些，并且让所有的事情在可控范围，一般采用过度的简化和宽泛的问题解决方式，这通

Such layouts are normally designed by hand: a complex and above all, time-consuming task that often has to be carried out on a trial- and- error basis. In order to make things easier, and to keep everything

在荷兰阿纳姆城市边缘的基地鸟瞰图。在这块 10 公顷的 "12 号地块" 上,我们的主要任务是使边界状况与当地居民将来生活环境相关的各方利益相平衡。

Bird's eye perspective of the building plot on the edge of the Dutch city of Arnhem. On the ten-hectare 'Field 12', the task was to balance the boundary conditions with the many and varied interests of the residents regarding their future living environment.

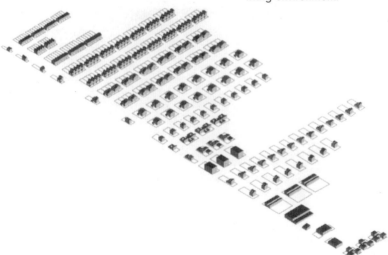

该图表展示将要被设置在场地上的各种类型建筑物的数量。每种建筑类型,不管独栋别墅、联排别墅或者一个虚拟的设施,都对场地有特殊的要求。

This diagram shows the number of buildings against all types of building which had to be arranged on the plot. Each building type, whether a villa, a terraced house, or a fabrication facility, places particular demands on the site.

任意设计 / Any Design

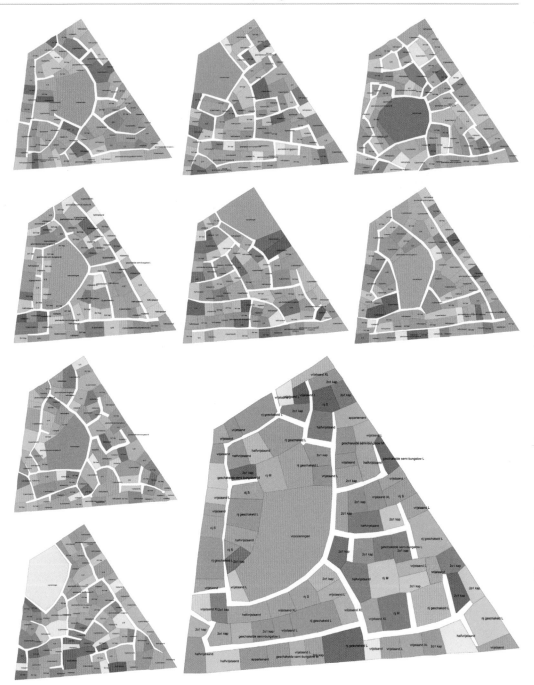

多层面考虑建设和评估的过程，为用地的划分生成基本布局形式，该布局同时考虑到未来居民的意愿。这种方法可以在很短的时间内生成一些变体，每一个都比之前的更加完善，并且我们不再用辛苦的手工方式来完成这个阶段的工作。

A multistage construction and evaluation process generated a layout for the distribution of the land parcels on the building plot, taking into account the wishes of the future residents, Instead of laboriously completing this stage by hand, possible variants were generated in a short time, compared to each other, and improved in a stepwise fashion.

常导致单一、僵化的发展模式。

然而，Schuytgraaf 项目采用了完全不同的方式。地块规划全部通过电脑生成——Kaisersrot CAAD 软件第一次在实际工程中应用。Kaisersrot 是德国凯泽斯劳滕大学和鹿特丹的建筑实践组织 KCAP 联合研究的成果（Kaisersrot 是将 Kaisersautern 与 Rotterdam 相结合的名称）。

Kaisersrot 依据信息科技领域中的既定原理，这些原理已经在实践中得到应用，包括粒子模拟算法、弹簧系统或者遗传算法。这些原理可进一步被采用和发展以适应建筑和城市规划的要求，Kaisersrot 为 Schuytgraaf 城市规划设计的机制是基于吸引与排斥的原理：在满足"Vinex"计划的要求下，用地上的每一个地块根据自身的特性（大小、位置等）都有自己特定的设计规则。Kaisersrot 所做的只是简单地让每一个地块同时"做自己应该做的事情"，与邻里地块进行竞争或合作。这个过程看起来很有趣，观察者能够以中立的姿态观察屏幕上的变化，看这些地块相互靠近或相互排斥，或变大、或缩小、或组团，直到最终形成一个整体。地块划分的编程借助了模拟算法，在这个算法中，它们最初都只是被设想成一个点。一旦它们的位置已经确定，则地块建

under control, strong simplifications and general problem-solving approaches are usually employed, which generally leads to developments taking on a uniformand rigid character.

However, Schuytgraaf was different. The development of the ground plan was carried out entirely on computer—the first real-world application of the Kaisersrot CAAD software. Kaisersrot is the result of a joint research project between the University of Kaiserslautern in Germany and the Rotterdam-based architecture practice Kees Christiaanse Architects and Planners (KCAP) (the name Kaisersrot is an amalgamation of Kaisersautern and Rotterdam).

Kaisersrot is based upon well-established principles in the field of information technology that have already found many applications in practice, including particle simulation, camera damping systems, or genetic algorithms. These were further adapted and developed to answer the demands of architecture and city planning. The mechanism by which Kaisersrot created the city plan for Schuytgraaf is based upon the principle of attraction and repulsion: the individual parcels of land were supplied with certain rules, according to their size and location, which conformed to the conditions stipulated in the Vinex program. The Kaisersrot program then simply allowed every parcel simultaneously to 'go about its own business' – either competing or cooperating with the neighboring parcels of land. This was a process which, to impartial observers at the computer screen, seemed to proceed in a somewhat playful manner as the plots of land attracted and repelled each other, grew or shrank in size,

设的具体功能将会被计算确定,用地界线也会随之确定。对于道路和人行步道,我们植入一套成熟的系统,确定每一个地块的最终朝向和比例。

Kaisersrot 展示了我们如何在工作中利用计算机的特定优势。计算机处理抽象过程如同数学推导过程,但执行大量命令时,速度要比人类快。Kaisersrot 在几分钟内生成的解决方案,借助人工处理将要耗费数周的时间。这种对于重复工作的提速效果,是逆转设计过程的一个有效的先决条件。通常情况下,城市规划项目将会使用"自上而下"的方法进行开发,先着眼于一般的、大规模的特征,再到特殊的、小规模的方面。但是,Kaisersrot 采用的是"自下而上"的方法:通过细节间相互的动态调整产生整体方案。

在传统工作流程中,像 Schuytgraaf 这样的地区设计方案,街道布局首先被确定,然后确定建设地块的布局,最后布局地块上的住房。Kaisersrot 的工作模式则恰好相反,房子是设计方案的"原始细胞",建筑地块由房子来定义,房子的位置决定了地块的布局。这种独特的

or formed new groups, finally coming together as a unit. The partitioning of the plots had been programmed with the help of a particle simulation algorithm in which the individual plots of land were, at first, represented simply by points. Once their positions had been determined, the contents of the respective building plots were then calculated and the borders set. For the roads and footpaths, well-known development systems were implemented, and these gave the individual building plots their final orientation and proportion.

Kaisersrot demonstrates how we can use the computer's specific strengths in our work. Computers process symbolic procedures in the same way they perform mathematical calculations, orders of magnitude faster than humans. Solutions generated in minutes by Kaisersrot would take weeks to process manually. This speed gain for repetitive tasks is the precondition for an effective reversal of the design process. Normally, city planning projects would be developed using a 'top-down' approach—that is to say, starting with general, large-scale features and working towards the particular, smaller-scale aspects of the design. Kaisersrot, however, uses a 'bottom-up' approach: the whole is a product generated by the mutual dynamics among the details.

In a conventional workflow, for an area such as Schuytgraaf, the layout of the streets would first be determined, then the building plots would be fitted in, and finally the dwellings on those plots. Kaisersrot works exactly in reverse: The houses are the 'primordial cells' of the design, according to which the building plots are defined and whose position determines the

此图显示了不断由演算重新生成设计方案。由建筑实践组织 KCAP 根据建筑和城市设计原则演示。

Here, the re-working of the algorithmically generated design is shown, as performed by the KCAP architectural practice, according to architectural and urban principles.

在 CAAD 实验室,我们的工作重心在于将实验转化成现实应用。

In the CAAD department, we place emphasis on the fact that our experiments can be turned into real-world applications.

方法，是人们手工处理无法实现的。

Kaisersrot 早期版本软件系统的评价体系仅限于本地，也就是说，只关系到它的近邻。一个单独的元素难以了解到整个系统的状况。此外，规划过程的各个阶段是一个连续过程并非同时进行。这些限制在后来的软件版本中都被消除了。

因此，这个过程的结果，即随后生成的地块的大小和比例并不能够与地块完全符合，也不能 100% 符合规划纲要，需要对细节作进一步完善。但是计算机提供了一个非常好的开端：早期的软件开发能够很好地找到各种不同需求之间的平衡。在首次的设计中，有说服力的形式语言和由此产生的有机结构一样令人吃惊，这样的有机结构在程序中无需任何索引。利用软件为 Schuytgraaf 生成的总平面，展示了结构在松散和有序间的平衡，就像我们经常所见到的历经时间慢慢发展起来的村庄模式。它看起来很古老，好像是有机成长起来的，而事实上，它是一个全新的电脑程序的产品。

layout of the plots. This approach is unique and is simply not achievable by hand.

A characteristic of this early version of the Kaisersrot software was that the evaluations carried out by the system could only be applied locally that is, only in relation to immediate neighbors: a single element would know nothing about the entire system. Furthermore, the various phases of the planning process would take place consecutively rather than simultaneously. These limitations were removed in later versions of the software.

Therefore, the results of this process namely the ensuing sizes and proportions of the building plots, did not conform of the building plots, did not conform 100% to the planning guidelines requiring further refinement of the details. But the computer provided a very good starting point: Already in this early phase of software development, the program proved to be outstanding at finding the happy medium between the various demands of the development. The convincing vocabulary of forms seen in this first design was as astounding as the resulting organic structure—without any indexes being required for that in the basic programming. The master plan for Schuytgraaf produced by the software demonstrated a balance between a very loose, and an ordered structure, which we often see in villages that have developed slowly over time. It looked old, as though grown organically, when it was, in fact, brand new and the product of a computer program.

我们还在鹿特丹 NAi 建筑展期间进行了新形式的民众参与活动。参观者可以自由地阐述他们对于建设地块位置的意愿和想法。不管地块是邻近周边，还是邻近森林或者公交车站，参观者都可以看到地块的实时更新，且并不破坏或扰乱整体方案的一致性。

相比以往任何时候，城市设计和建筑设计都需要满足更多的要求，像 Kaisersrot 这样的软件给设计过程带来了透明性，并且使其与真实世界的互动关系得以呈现和确定。这种透明度能够帮助人们对设计的结果进行理解和交流。这使设计过程中的"自上而下"改为"自下而上"，使设计过程焕发了新生，并使公众和最终业主得以尽早参与到开发过程中。此外，在解决问题方面，该软件不仅为通用的城市设计绘制出粗略的设计示意图，也能够使单个项目内容处于可控状态。总之，这个计算机辅助规划的过程给了我们更多自由，使我们可以将精力投入其他方面。例如在设计过程中，不会失去对结构的整体把控，也不会忽略项目的实际问题。

We also implemented a new form of civic participation during an architecture exhibition at the NAi (Nederlands Architektuurinstituut) in Rotterdam. Visitors were able to express their wishes and their ideas about the locations of their building plots—whether they were next to their favorite neighbor or close to the forest or to the bus stop—and see them update in real time, without breaking or disturbing the coherence of the overall plan.

More than ever before, urban design and architecture need to be able to meet myriad demands. Software like Kaisersrot lends a transparency to the process and enables interrelationships with their real-world interplay to be shown and identified. This transparency helps to make the consequences of individual choices comprehensible and negotiable. The change in design process from 'top-down' to 'bottom-up' gives a new lease of life, and allows the public and end users to be involved much earlier in the development process. Additionally, in its suggestions for solutions, the software is not limited to rough schematics of general urban design principles: It also takes the individual project context as a controlling situation. Altogether, this type of computer-aided planning process gives us more freedom to invest our capacities in other aspects, for example the design, without losing sight of the structural and pragmatic aspects of the project.

项　目：**Heerhugowaard（荷兰）**

时　间：2005—2008

参与者：Markus Braach, Oliver Fritz, Alexander Lehnerer

合作者：Kares en Brands (Hilversum, NL);
　　　　Kees Christiaanse Architects and Planners (KCAP) (Rotterdam, NL)

发展性规划设计　Arranging Historical Growth

Heerhugowaard 项目位于荷兰北部，其要求与 Schuytgraaf 大致相同，即设计和规划一个新的建设开发区。二者的主要区别是，Heerhugowaard 的建设规模更大，地块、房屋、住宅的数量是 Schuytgraaf 项目的五倍以上，总面积将近 70 公顷。与 Schuytgraaf 项目相比，不仅项目内容的需求增加了，软件也进一步升级，使其更为强大。

正如荷兰大部分地区的情况，小区建设的最主要考虑因素是地下水的状况和土壤条件。基于当时的地质条件，该地区被分为三个主要区域，每个区域中建造特定属性的建筑物。这导致各小区的建设都有特殊的规则，位置、大小及比例都要进行计算。

Kaisersrot 软件关键的进步是它增强了活力。首先，它可以同时并行地处理多个问题。这很适用于该项任务，并能不断地给出更好的结果。为了说明这一点，举一个具体的例子：

The requirements for the Heerhugowaard project in northern Holland are broadly similar to those for Schuytgraaf, i.e., the layout and planning of a newly built development. The main difference is that it was to be laid out over a considerably larger area, with more than five times as many building plots, houses, and apartments encompassing a total area of almost 70 hectares. Compared with the Schuytgraaf project, not only had the demands of the project context grown, but the software itself had been further developed, and was considerably more powerful.

Key considerations for the building plot were—as they so often are in Holland—the groundwater situation and the varying soil conditions. Based on the prevailing ground conditions, the area was divided into three main zones on which only buildings with certain properties could be built, or, in some cases, not built. This resulted in addition　rules for each building plot, for which not only position and size, but also proportions had to be calculated.

The key to progress was in the increased dynamism of the Kaisersrot software so that, firstly, it could process the individual instances simultaneously and in parallel. This suited the task and consequently gave much better results. To illustrate this with a concrete example:

在70公顷的面积内,随着分化程度需要灵活地规划,不可能用传统的方法实现。

Given the area of 70 hectares, it is obvious that flexible planning with the required degree of differentiation would not be possible using traditional methods.

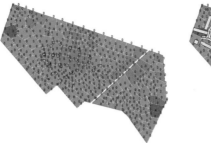 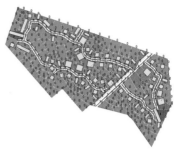

在第一步中,小地块被安排在外围。不同颜色代表不同的土壤条件。合适的建筑类型从一个可用的数字目录中选择,包括建筑地块的尺寸、土壤条件和之前确定的道路系统。

In the first step the parcels of land were arranged around the perimeter. The colors represent the varying soil conditions. Suitable building types were chosen from an available digital catalog, subject to the size of the building plot, soil conditions, and previously defined main circulation.

在此基础上,下一步我们看见建筑用地之间形成了次一级的道路系统。最后的布局符合了居民的个人意愿以及建筑师的思想,形成了貌似从原始的、有机生长发展而来的形态(就像一个村落的布局)。

Using this as a basis, the next step saw the buildable plots worked out with respect to a secondary circulation system. The final layout corresponds to the individual wishes of residents, as well as to the ideas of the architect, and forms a prototypical, historically grown settlement pattern. (like that of a village)

以上三个例子展示了在成功确认参数后,变量生成不同的场景。街道布局、地块大小和建筑物的位置都遵循相同的规则。每个方案都展现出不同的城市设计构型。

Three examples of variants demonstrate how, after successful validation of the parameterization, various scenarios play out. The layout of thoroughfares, size of plots, and location of buildings all follow the same set of rules. However, each variant expresses itself through differing urban design configurations.

建设地块的位置、大小、比例及街道和道路的布局彼此联系密切、动态变化。这些紧密结合、相互依存的关系带来一个循环模式，Kaisersrot 能够使用递归、循环优化算法来处理这种循环模式，并对 Heerhugowaard 未来的发展进行计算。其次，Kaisersrot 采用遗传算法来生成最初建筑用地的划分方案和后续优化。简单地说，它可以随机生成解决方案，然后进一步完善那些最优秀的方案。还有一点与 Schuytgraaf 项目的差异是，计算结果并不只是注重某个地块和它周边邻居的情况，而是从全局着眼，进行整个系统层面的运算和优化。

像 Heerhugowaard 这样的大型项目一般要在几十年的时间内经过不同的阶段逐步完成，同时历经不断地修正。无论是对建设成本或住房数量，Kaisersrot 软件都使我们能够着眼于发展的角度，且不失去对整体的把控。这个规划可以根据过程中的变化不断地进行自我调整，如果有新的变化发生或加入一些新的布局安排，规划也无需进行新的设计和返工。在城市规划这个层面的设计一般都是一个动态的过程，设计的结果最好也只是临时性的结果。随着计算机辅助设计的支持，这个过程可以得到更好的实现，同时具备更细的分化程度和更高的效率。这是结构式程序设计的优点在于：在不同解决方案的具体操作形式中，从开始到任意选择的一个终止点，"结果"始终保持灵活性。

The position, size, and proportions of a building plot, and the layout of streets and roads depend upon each other in a dynamic fashion. These tightly bound interdependencies give rise to a circulation pattern that, for the Heerhugowaard development, Kaisersrot was able to work upon using a recursive, circular optimization algorithm. Secondly, genetic algorithms were used for the primary partitioning and optimization of the building plots, which, simply put, generate random solutions and then further refine only those that work best. The final difference from the Schuytgraaf project was that the results of the calculations were not just geared to, and evaluated at, the local level—in conjunction with the neighbors—but also at a global, system-wide level.

Large-scale projects like Heerhugowaard are generally completed in various stages over decades, all the while continuously being revised. The Kaisersrot software gives us the ability to start in one corner of the development without losing sight of he overall picture, whether it be building costs or the number of dwellings. The plan adapts continuously with the changes that happen during the process, but does not have to be totally redesigned and reworked if a few changes are made, or if new framework conditions or new layouts have to be taken into account. Urban planning at its broadest scale is always a dynamic process that produces, at best, only provisional results. With the support of the computer, this process can be better utilized with considerably better differentiation and efficiency. This is the specific strength of structural programming: the 'results'—in the form of concrete proposals for various possible solutions —always remain flexible, from the beginning to any arbitrarily chosen end point.

项　目：Oqyana（阿联酋，迪拜）

时　间：2006

参与者：Markus Braach, Alexander Lehnerer

合作者：OQYANA World First (Dubai, UAE); Institut für Bildbearbeitung,
ETH Zürich (CH) - Pascal Müller, Simon Haegler;
Professur für Gebäudetechnik, ETH Zürich (CH) - Prof. Hansjürg Leibundgut

把形式交给数据 Giving Form to Data

过去几年，全球范围内建筑项目如雨后春笋般拔地而起，不断挑战着欧洲——特别是瑞士——有关规模的理念。

"Oqyana 世界第一项目"就是巨大工程"世界岛"的一个组成部分。在波斯湾，正在修建以澳洲大陆为形状的 20 座人工岛。这些岛上，共建有 2 000 多处住房、别墅、公寓，可容纳约 12 000 人。工程造价将达到 35 亿美元左右，2012 年竣工。

在很多方面，工程项目都对现有规划步骤提出极大的挑战。一方面项目的规模在不断加大，另一方面需要满足个体客户针对他们自身需求所提出的要求。所有这些不同的要求须整合到一个统一、稳定的总规划中，所有的工作必须限期完成。因此，就得把城市规划的需求、建筑层面的供水要求、技术基础设施、各种各样的客户要求与总体建筑质量以及时间限制整合起来。初步估计，这一工程需要约 40 家专业承包商的通力合作。然而，如果他们都按照以往的工作方法各自去施工的话，工程的难度会变得非常大。就算经过一年的精心规划，期间并无任何赘余的开支，他们之

In the last few years, building projects have sprung up worldwide that defy European- and especially Swiss-ideas of scale. 'Oqyana World First' is part of one such megaproject named 'The World.' Twenty artificial islands are being created in the Persian Gulf, in the shape of the continent of Australia. On these islands, some 2000 houses, villas, and apartments are being constructed to accommodate approximately 12,000 people. The cost of the project— scheduled to be completed in 2012—is around $3.5 billion.

In many ways, the project poses special challenges for the planning process. In addition to the sheer scale of the project, there is a clientele that sets great store by individual consultation and planning tailored to their needs. All these differing demands must be integrated into a consistent and stable master plan, all the time working to a tight deadline. Therefore, marketing has to integrate the needs of city planning and architectural vision with the demands of water supply, technical infrastructure with the many and varied demands from individual clients combined with overall construction

间的合作也很难令人满意，因为仍有工程造价超支 40% 的风险。如此大规模的工程需要面对各种各样且往往各自矛盾的需求，但对我们的 Kaisersrot "引擎"来说，这正是它发挥强大解决问题能力的良好平台。

我们的研究任务主要集中在工程项目相对较早的阶段。根据设定好的人工岛的规模和城市发展规划的概念，确定该工程项目的基本参数，从而在进一步的规划过程中，不会增加技术方面和资金方面的风险。所以，在开始阶段，我们关注的不是固定的建筑理念，而是一种工具，这种工具能在城市规划框架内统一整合未来预期客户意愿的数据，并通过视觉画面的方式，使其交互作用更一目了然。

借助我们的程序，我们能快速回答关于开发过程中的所有基本问题，也能根据各种要求产生丰富多样的可行性解决方案。比如，建筑在人工岛上的布局如何？有多少建筑能直接面对大海？需要修多少防波堤？整个开发是如何进行的？各个人工岛的基础设施是什么？整体竣工情况又如何？建筑高度与密度的关系是怎样的？风景的分布与视线对整个系统有何影响？如

quality and time constraints. At a rough estimate, this requires the collaboration of around 40 specialist contractors who, if they were set in their normal working practices, would soon be swamped by the task. Even after a year of intensive planning and no small financial outlay, the orchestration of these contractors has hardly been satisfactory, so that there still was a risk of a 40% financial overrun. Projects such as these, with their varied and often contradictory requirements, provide the ideal stage on which our Kaisersrot 'engine' can display its special problem-solving capabilities.

The main focus of our study centered on an earlier phase of the project. Setting the size of the islands and developing an urban planning concept should fix the fundamental parameters of the project, so that further planning can be carried out without additional technical or financial risks. Therefore, initially it was not so much about fixed architectural ideas, but rather about a tool that is capable of integrating future data of the expected client wishes coherently within an urban planning framework and then making its interplay comprehensible by displaying it visually.

With the help of our programming, we were able to quickly answer the fundamental questions about the development and to generate a rich diversity of possible solutions based on a variety of requirements. What should the distribution of buildings on the islands look like? How many buildings should have direct access to the sea? How many jetties would be needed? What should the development look like? The infrastructure of the

果有客户想要欣赏到完整连续的海景，那么造价是多少？

使用传统的方法，对这些复杂的问题提出一套有效的解答并给出可行的方案，会显得捉襟见肘；如果要提供多种不同版本的新方案、新规划，在不耗费巨大的人工和时间成本的情况下，绝对是无法实现。然而，对于运行相关算法的电脑来说，这却是非常轻松的事情，它可以产生各个层次的连续结果，从最小的细节到总平面图中建筑的编号、位置、高度以及成本、收入的计算。这些数据可以以 Excel 表格或者图形的形式给出，也可以用三维数据建筑的形式给出，这只需要你选择相应的输出格式即可。像 Kaisersrot 这样的软件的好处就在于处理和组织数据，电脑可快速并无误地运行，从而可使建筑开发和城市规划达到新的整合高度，自动地得出适用并满足各种不同要求的解决方案。

每一个建筑方案，即使像"世界岛"这样规模的方案，同样是由许多利益方来决定的。工程师负责技术，建筑师负责设计，城市规划师负责制定总

individual islands? The entire finished product? What should the relationship between building height and building density be? What effect do the provision of vistas and lines of sight have on the whole system? How much would it cost if a client were to want an uninterrupted view of the ocean?

It is already difficult enough, using classical methods, to provide one valid answer to these complex questions and to come up with a workable proposal, but it would be well-nigh impossible to deliver new plans with new proposals without an enormous cost in labor and time. However, for an appropriately programmed computer, this is almost a trivial exercise, which produces consistent results at all levels, from the smallest detail to the overall plan: from the number, position and height of the buildings through to cost and revenue calculations. Whether these numbers are provided in the form of Excel spreadsheets or in the form of a diagram, a quasi three-dimensional data architecture, is simply a matter of choosing the appropriate format.The new quality that can be achieved with a piece of software like Kaisersrot is a matter of handling and organizing the data so that the computer can process it much more quickly and without errors. In this way, architecture and urban planning can reach a new level of abstraction, from which appropriate and individual solutions arise almost automatically.

Every building proposal—even more one on the scale of 'The World'—is shaped by the various interests and approaches of its stakeholders. The engineers are responsible

Oqyana (Dubai, UAE)

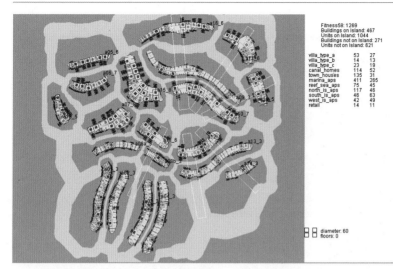

迪拜近海的 20 个人工岛，构成微缩的澳洲大陆形状。诸多不同类型的建筑单体，以不同的颜色表示，在岛上合理排列，以避免彼此冲突。不同的设计方案按照城市规划设计原则，可进行空间、经济等方面的比较。

Off the coast of Dubai, 20 artificial islands form the shape of the continent of Australia in miniature. A large number of individual buildings of varying types—represented here by different colors—had to be arranged on these islands in order to avoid conflicts. The alternatives could be compared spatially, economically, and according to urban design principles.

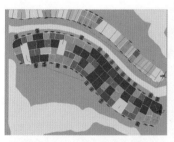 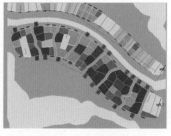 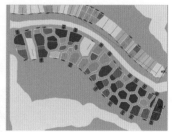

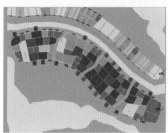 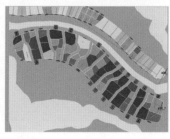

建筑参数的变化，比如让建筑更加靠近海边，可立刻计算出并在整体的布局图中体现出来。

A change in the parameters—like moving the buildings closer to the water's edge, for example—is calculated immediately and reflected in the layout of the entire settlement.

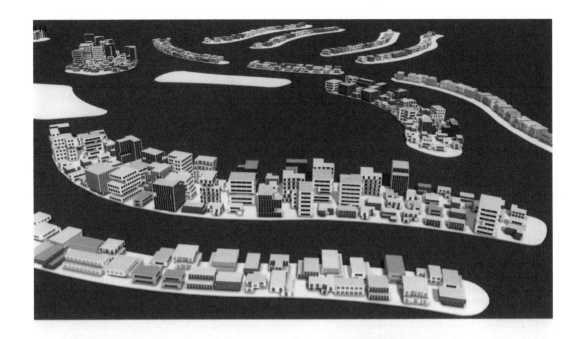

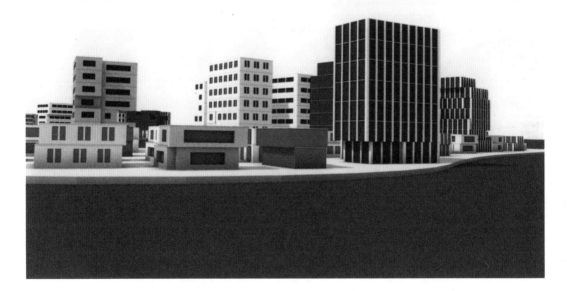

在该研究中，许多参数（如高度、密度、建筑拓扑、居住人数、阴影研究、建筑功用、近水情况等）都是相互影响的。对这些参数的多种考量会产生不同的布局，可在视觉上直接进行比较。

In this study, many parameters (heights, densities, building typologies, number of inhabitants, shadow studies, mix of use, access to the water, etc.) were played off against one another. Varying weightings for these parameters led to different layouts that can be directly compared visually.

体布局，市场专家负责销售，而经济师则负责制定和管理预算。不难想象，这些不同利益方，如果只顾推进各自的进度，会给整体项目工程建设带来巨大压力。

在计算机的帮助下，将各个群体的不同利益进行一一映射，可以使问题变得更容易控制：尽管这是一个自动的过程，但仍可保持解决方案的多样性。大至对整个系统或小至对不同细节的要求，都可以通过一目了然的方式呈现出来，所以在各领域专家通力合作制定的框架下，这些相互制约的要求可以被快速有效地权衡处理。

当然，市场条件或者客户要求一旦更改，这些规则也会迅速调整。这种在整体连续系统中的灵活性，可以让我们更大程度上降低那些通常工期长、风险大的工程的风险性。尤其是当你接手的是一个35亿美元的项目，这意味着成本的风险性更大。

for technical aspects, the architects for design. The urban planners work on overall layout, the marketing experts take care of sales, while economists draw up and manage the budget. It is not hard to imagine that these differing interests can cause tension within the project, with each party pushing their own agenda.

The mapping of such diverging interests in multifocal constellations using computer-aided optimization makes the problem considerably more manageable: In spite of being an automated process, it maintains the richness of many possible solutions. The demands placed upon the entire system and the diverse and numerous details can be brought together in a transparent relationship, so that—under the framework conditions and rules laid down by the various experts in their fields—the many competing demands can be balanced easily and efficiently.

The rules can, of course, quickly be adjusted when market conditions or client demands change. This flexibility within an overall consistent system allows us to considerably reduce the risks of the often long-drawn-out and roller-coaster-like projects. And when you are dealing with a $3.5 billion project, that can mean a lot of money.

项 目：**Globus Provisorium**（瑞士，苏黎世）

时 间：2004

参与者：Markus Braach, Oliver Fritz, Alexander Lehnerer, Christoph Schindler, Philipp Schaerer

合作者：Tobler & Partner (Basel, CH)

赞助商：Hochparterre – Verlag für Architektur und Design (Zürich, CH)

共识引擎 The Consensus Engine

苏黎世中部的 Papierwerd 区，位于中央火车站附近的 Limmat 河岸，是该城最著名也最具有争议的开发地点。无数的更新或者取代现有的 Globus Provisorium 的尝试均以失败告终，因为该项目相关方面的利益冲突太大了。这是建筑工程难以取悦所有人的经典例子。不管提出什么样的方案，总会带来问题：不是阿尔卑斯山壮丽的风景被破坏了，就是阻断了美丽的 Limmat 河岸，要不然就是离中央车站太远了。反正总有人对结果不满意。

The Papierwerd district in central Zurich, which lies directly on the Limmat River near the central railway station, is one of the most prominent—and most controversial—development sites in the city. Countless attempts to renew and replace the existing Globus Provisorium have failed under the weight of the conflicting interests of the parties concerned—a classic example of an architecture project in which it was impossible to please everybody. Whatever was proposed brought its own problems: either the splendid view of the Alps was ruined, or it blocked the wonderful Limmat riverbank, or it was too far away from the main railway station. There was always someone who was unhappy with the result.

当 CAAD 实验室参与到这一场地多功能建筑的竞争中，我们并无意从传统意义上提交确定的设计方案。对于我们来说，我们要做的只是开发一系列规则，作为制定一份站得住脚的设计方案的前提条件，以适应当时棘手和麻烦的状况。因此，我们的竞赛方案就有些挑战的意味。而其他建筑师展示的是他们已经完成的模型，并非常自信地认为他们的方案才是所有竞争 Globus Provisorium 项目中最

When the CAAD department took part in a competition for a multi-use building on this site, it was, therefore, never our intention to deliver a finished design in the classical sense. For us, it was more about the development of a body of rules as a precondition for drawing up a tenable design in accordance with the rigid and problematic framework conditions. For this reason, the presentation of our competition

Globus Provisorium (Zurich, CH)

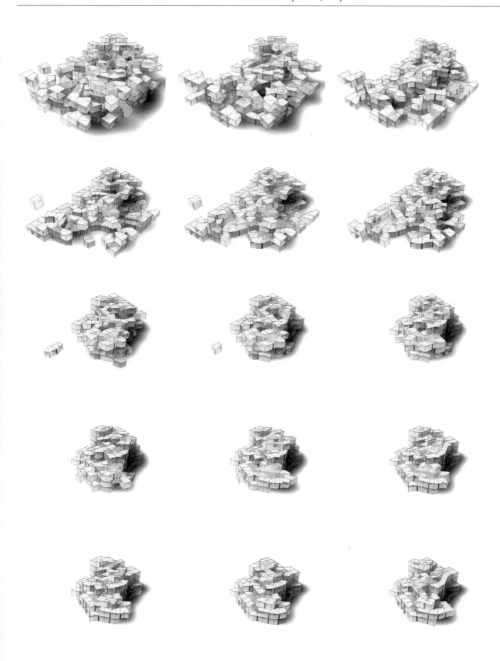

在苏黎世中心 Limmat 河岸的 Globus Provisorium 项目竞争中，我们一开始就表示，建筑问题可以结合城市设计的问题一并解决。不同用途的空间用不同的颜色区别，如居住区、酒店或者商业区。

With the competition for the Globus Provisorium, on the banks of the River Limmat in the center of Zurich, we were able to show for the first time that architectural problems could be engaged alongside those of urban design. The colors represent the various uses of space, such as residential, hotel, or retail.

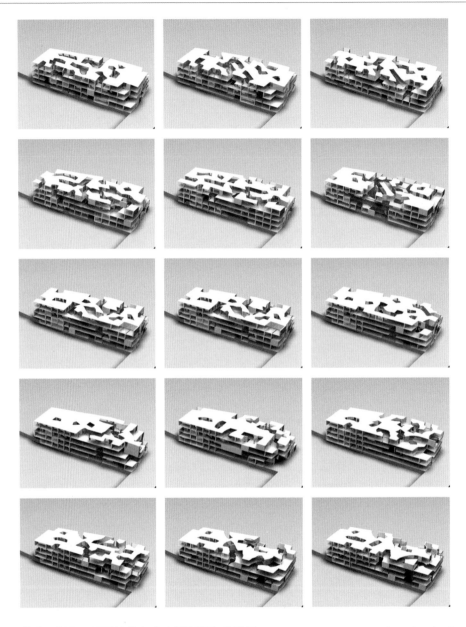

首先是根据不同用途将各个房间用颜色代码区分，在建筑模拟中以随机的方式进行排列布局。在计算过程中，各房间的位置会不断变化，从而符合或者改进与周围房间、景观或者路线长度的关系。借助演化算法，通过多次重组生成，从而产生一个布局方案，使得每一个房间的位置在几何和拓扑上表现堪称完美，以至于任何进一步的局部修改都将造成宏观上布局的明显缺陷。

The room layout, with individual rooms color-coded according to use, was initially arranged in the building shell in a random fashion. During the course of calculation, individual rooms changed their positions in order to conform to or improve upon relationships with neighbors, views, or the lengths of routes. With the help of an evolutionary algorithm, after many permutating generations a layout was reached in which each room had a geometric and topological position that conformed so well to the criteria that any further local changes would lead to an obvious deterioration of this layout on a global scale.

好的解决方案。而我们并没有提供一个固定的方案。准确地说，我们提交的是一份灵活性的解决方案，针对最终建筑位置和建筑本质属性的一些问题：建筑体是应该高还是低，是"胖"呢，还是"瘦"，说实话，我们也不知道。但我们很清楚我们能在任何时候根据任何指导大纲做出一份可行的解决方案，可能会正合适。该项目将充分展示这一点。

初始情形如下：7600平方米的地块上应包括餐馆、超市、宾馆，每个都有自己特定的要求：体现在地点、基础设施以及在该复合体中空间和店铺群的发展和设置方面。外部最大的挑战在于如何把这一项目与苏黎世城的风格统一起来。仅有的参数就是建筑场地的外界线和可能的最大建筑高度。

我们工作的中心是，设计一个电脑模型，相当于一个"共识引擎"工具，整合所有股东的利益，尤其是工程的主要股东——城市政府和投资人。该城市一直关注这一区域的旅游潜力，也就是说，湖边景色、教堂以及阿尔卑斯山风景不应该被建筑阻挡，主要景观不能被破坏。而另一方面，投资人却更加在意功能的最佳配置，建筑所在的位置能使它们发

entry had the potential to be somewhat provocative. While other architects had presented finished models, confident that theirs represented the best of all possible solutions, we presented no fixed proposal. Rather, we submitted a flexible solution to questions concerning the final location and architectural nature of the building: should it be tall and compact, or low and wide? We simply didn't know. What we did know, however, was that we could offer a functional solution at any time, according to any guidelines—no more, no less. The project would serve as a demonstration of this capability.

The initial situation was the following: a plot of some 7600m² should contain, among other things, a restaurant, a supermarket, and a hotel, each placing their own demands on the plot, on infrastructure, and on development and placement of the spaces and shop units within the complex. Externally, the greatest challenge was the integration into Zurich's city fabric. The only parameters were the external boundaries of the potential construction site and the maximum possible building height.

The focus of our work was the generation of a computer model that would function as a sort of 'consensus engine,' bringing together the interests of all stakeholders, especially the main players in the project—the city and the investors. The city had always been concerned about the tourist potential of the area. This meant, therefore, that the views

挥最大效益，然而这样的做法不可避免地会直接影响到湖景的观赏和中央火车站的接近以及内部组织。鉴于这些因素，一个异常复杂的相互依赖和参考的互动问题就产生了，这不是简单的人工操作就能解决的——至少，在有限的时间内无法解决。当然，任何问题如果单独分离出来都能解决，但在相互影响的错综复杂的过程中，在经常矛盾的参数和规定情况下，就难以解决了。

我们编写的模拟模型完成了两个主要任务：第一，发展潜力较高或者较低的地域按照它的视野好坏或者场地发展潜力确定。第二，空间布局在建筑外围结构（建筑体量的限制）内进行分配，或者与外部环境形成理想关系（如湖、火车站等），或者与周边空间形成理想关系。所有投资人都会对项目产生直接的影响。一旦拨动了"转盘"（规则的定义和权重），一个新的结果就产生了。因此，很明显，这不会产生一个确切的设计方案，而是为设计方案形成基本规则，并在产生让各方所接受的共识中体现其价值。

to the lake, the churches, and the Alps should remain uninterrupted, that the main vistas / views should not be obscured. Investors, on the other hand, were more interested in an optimal mix of uses, that the spaces be positioned so that they function optimally, which, by implication, has a direct influence upon the view to the lake, the access to the main train station, and the internal organization. Because of these factors, an extremely complex interplay of interdependencies and references raises that simply could not be managed manually—at least, not within a justifiable time frame. Of course, any problem can be solved when tackled in isolation, but not within a networked process defined by other—and often conflicting—parameters and rules.

The simulation model that we programmed fulfilled two main tasks: Firstly, areas with higher or lower development potential were identified on the basis of sight-lines or site development potential. Secondly, a layout of spaces was distributed within the building envelope (the limits of the volume of the building) that traded off, on the one hand, the ideal relationship to the environment (lake, railway station, etc.), and on the other hand, the ideal relationship with the surrounding spaces. All the stakeholders could have a direct influence on the project. As soon as one 'turned the dials' (the definition and the weighting of the rules), a new result was produced. It was, therefore, clearly not a case of developing an explicit design proposal, but rather to first develop the ground rules for a design, which would show its worth by generating an acceptable consensus for all concerned.

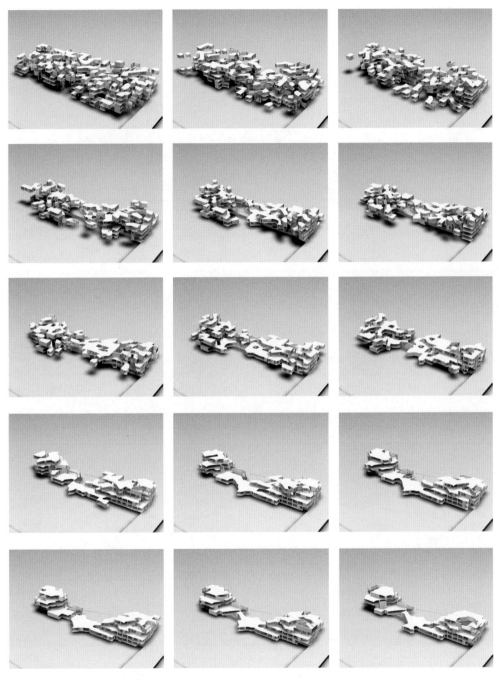

与城市规划问题一起，该项目内部两个主要功用的空间，酒店（蓝色）和餐馆（红色）需要再三考虑：酒店房间需要面对湖景，商场（绿色）需要尽量放在地面层，并靠近火车站，餐馆应尽量靠近酒店大厅以及商场的活跃空间。

Alongside urban planning questions, the internal interests of the two main uses—a hotel (blue) and a restaurant (red)—want consideration: hotel rooms require a view to the lake, markets (green) need to be on ground level and close to the railway station, and the restaurant should lie as close as possible to the hotel lobby and the activity in the markets.

从 Limmat 河岸目测三个变化形态。这一项目展示的是房间布局与建筑外围结构的不同组合是如何被测试和比较的。通过这一方式，在做建筑决策时风险会更小。

Visualization of the three variants, as seen from the bank of the Limmat. The project demonstrated how different combinations of room layouts and building shells could be tested and compared to each other. In this manner, building decisions could be taken with less risk.

从技术角度来看，演化算法软件通过连接单个模块与动态几何图形，来计算单个空间的分布与排列，从而使空间的形状和大小最优化。整个系统发展的起点是一个建筑单元。这些建筑单元根据设定好的任意条件，集合成一个建筑整体。一个非常有趣的现象是，每个建筑单元首先又是由它所处的环境来定义的：应该是建筑单元呢，还是流通单元呢？是否需要门或窗？如此，数据结构会很自然地呈等级排列出来：空间是由一组单元组成，建筑由一组空间组成，以此类推。因此，我们得以获得更多的回旋余地，在颇为抽象的层次上、在 Globus Provisorium 项目中体现得更为明显。各元素自由组合，可以更清楚地产生各种空间上的布局，这将是建筑设计上的一个令人兴奋和心动的新起点。

From a technical perspective, the software combined an evolutionary algorithm that calculates the distribution and arrangement of the individual spaces by swapping individual modules with dynamic geometries that are responsible for the optimization of the form and the size of the spaces. The starting point for the development of the whole system is an individual cell that grows into an entire building, according to any conditions that one may wish to set. One of the most exciting aspects is that each cell is defined first of all by its context: Will it be a building cell or a circulation cell? Will it need doors and windows? In and windows? In this way, a natural hierarchization of data structures emerges: A space is a group of cells, a building a group of spaces, and so on. Thus, we were able to gain a kind of formal leeway, which, on a rather abstract level, becomes obvious in the Globus Provisorium project. The freedom of the elements to organize themselves clearly leads to spatial layouts that could provide an exciting and appealing starting point for the design work of the architect.

项　目：**Hardturm（瑞士，苏黎世）**

时　间：2006—2007

校　订：2007 (Block B)

参与者：Markus Braach, Benjamin Dillenburger, Philipp Dohmen, Pia Fricker, Alexander Lehnerer, Steffen Lemmerzahl, Kai Rüdenauer

合作、主办方：Halter Unternehmungen (Zürich, CH)

数字链 The Digital Chain

我们为瑞士企业 Halter 所展开的这个研究项目的特别之处在于，从规划直到生产过程中对细节深度的把握和坚持使用尖端前沿技术。这项研究最重要的方面就是第一次采用我们的方法建造出一座实体建筑：在苏黎世郊区的 Hardturm 建造一处公寓街区，拥有 300 幢寓所，建造预算 2 亿瑞士法郎。

为这个方案开发的数字链控制系统包括三个基本阶段：首先，通过这种系统的优化处理，使基于自动生成的建构框架可以满足不同形式的使用要求，这些要求基于各种可接受的参数形成。其次，建立汇集优质平面布局图的数据库系统，这些平面布局可以匹配多种特殊的使用要求，从而方便客户的独立选取以及与原始建构框架的整合。再次，设计一种特殊的流水线式建筑立面建造方法，可以让立面本身承载大部分服务技术，这就意味着立面可以迅速适应不同的建构理念。

对于第一阶段，我们的研究任务简单说来就是尽可能地按照建筑规则

The special features of this study, which we carried out for the Swiss firm Halter Unternehmungen, were the depth of detail and the consistent use of cutting-edge information technology from planning right through to the production process. The most important aspect was that this was the first time that a real building had to be developed: an apartment block in the Zurich suburb of Hardturm, with around 300 dwellings and a construction budget of 200 million Swiss francs.

The digital chain of command developed for this project consisted of three fundamental phases: Firstly, a mainly automatically generated building framework that was optimized in such a way that it could best accommodate different forms of use based on a variety of accepted parameters. Secondly, a library of good ground plans suited to particular uses, from which the clients could choose individually, to be integrated into the raw building framework. And thirdly, a special method of production-line facade construction whereby the facade itself contained large parts of the service technology —which meant that it could be quickly adapted to the various architectonic ideas.

In terms of the first step, more simply put, it was a matter of arranging various uses with regard

来安排各种使用功能。更具体地说，对于投资者和未来的居住者，这个过程就是关于如何充分利用建筑用地——在这个案例中即为实现住房建造数量最大化的同时保证个体公寓质量的最优化。

整个项目的一个特征是计算机处理系统与规划者的要求和想法变化之间的关系。建筑的边界由建筑师和城市规划师设定，而后计算机根据特定的规则在设定区域内填入建筑最大体量的外围结构。这种"鸡尾酒"式开发模式——即它的各部分中可以被填入各种不同的功能——是由规划者预先设定的。所以，当我们讨论"功能"时，我们的脑海中并没有浮现出严格定义上的平面图，而是一种粗略定义的类型——例如，"120平方米的公寓住宅"或者"300平方米的零售商店"。在预设界限内，各种功能的分布和调整是由计算机来完成的，但是仍然要考虑到预设的评价标准，例如景观、噪声影响和自然光的获取。这种结果最终按照不同使用功能类型和相应的规模、发展情况、日照及噪音影响的要求对"鸡尾酒"进一步调整。在最后一步时，借助于进化算法，软件可以将各部分建筑单元拼合在一起，这样各条标准得以更可能地被满足。

那么完美的解决方案应该是怎样的呢？尽管这么问不是很好，但方案结果是应该经受得起客观的分析、测试和比较。并不是说计算机比建筑师或规划师在住宅设计上更加出

to building regulations as well as we possibly could. Put more concretely, for both investors and future residents, it was a matter of making the best use of the building plot—which in this case meant securing the largest possible number of dwellings, but at the same time guaranteeing the highest possible quality for the individual apartments.

A characteristic of the whole project is the relationship between computer-based processing and the requirements and changes made by the planners. The boundaries of the building are defined by architects and urban planners. The computer then fills out this envelope to its maximum building volume under compliance with certain rules. This 'cocktail'—i.e., which volumes can be filled with what uses—is predefined by the planners. So, when we talk about 'uses', there is no rigidly defined ground plan in mind, but rather a roughly defined typology-like, for instance, '120m^2 apartment' or '300m^2 retail premises.' The distribution and ordering of the various uses within the predefined limits is carried out by the computer, but still with regard to predefined assessment criteria, such as views, noise exposure, and access to daylight. The results serve eventually to further differ-entiate the 'cocktail' based on the types of uses and the corresponding demands on size, development, access to light or exposure to noise. In a final step, and with the help of evolutionary algorithms, the software puts the building blocks together so that the individual criteria are fulfilled as successfully as is possible.

So what does the perfect solution look like? Although this might be the wrong question to ask, the solution is amenable to objective analysis , measurable, and able to bear comparison. It's not a question of whether the computer can design

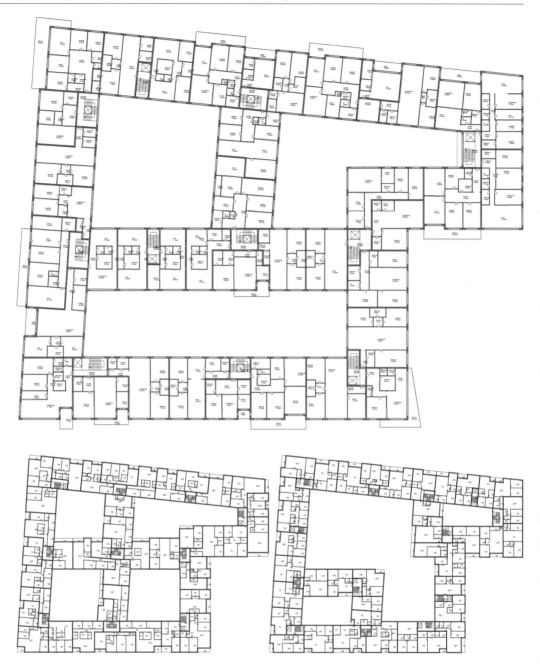

在程序早期计算阶段的 Hardturm 开发地区平面图。

Ground plan of the development area of the Hardturm project in an earlier phase of calculations.

Hardturm (Zürich, CH)

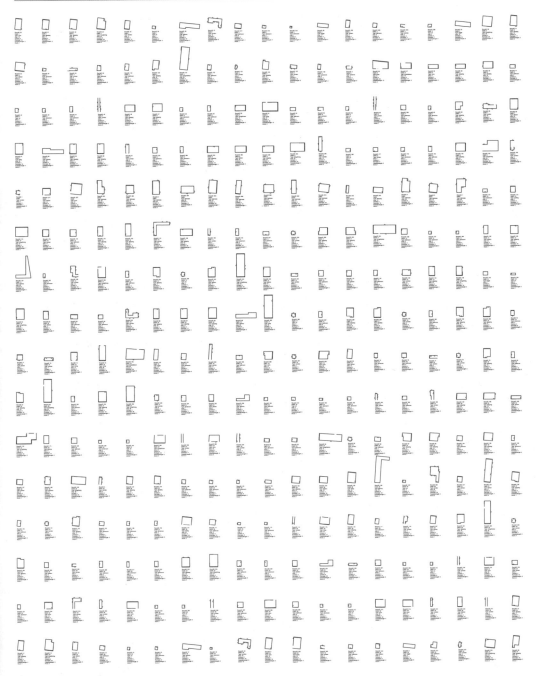

在多级选择和调整程序中，对苏黎世过去20年内新建建筑综合样本的平面图不断进行调整，以适应现存的建筑框架。这个研究成果是一种自动生成的不同平面布局的拼合体，它充分利用现有的空间条件（规模、位置、日照、噪音、景观和路径等），创造出一种理想的多样化住宅规模和使用功能的混合体。

Ground plans from a comprehensive catalog of new buildings in Zurich over the last 20 years were adjusted to the existing building shells in a multistage selection and adjustment process. The result was an automatically generated patchwork of different ground plans making the best use of the existing spatial conditions (size, position, sun, noise, view, connecting pathways, etc), and producing the desired mix of varying sizes of residences and mix of uses.

色：关键在于，无论评价的标准是每平方米的造价或是日照量，计算机都能迅速给出很多种解决方案，并对它们进行比较得出最终的评价结果。某些定性的标准，例如噪音量或是光照度，都可以利用这些程序被最终确定和量化处理。

这种大型项目的一个明确的特点是平面图个体特征的研究——这也是我们在数字链系统研究中的第二阶段。现如今，个性代表了一种与众不同的人类需求，尤其是在讨论私人住宅问题时。每个人都希望拥有一种个性化的，似乎是为其个人量身定制的环境氛围。但在例如 Hardturm 的大型住宅区开发案例中，想要在传统的规划设计约束条件下，实现这一点几乎是不可能的，因为在一种理性的资源投资范围内，各个独立的个性化平面布局方案在建造或是规划阶段就很难被实现。

针对 Hardturm 研究案例，我们开发出一种可以提供个性化的建筑平面布局的程序，它代表了一个逻辑步骤，可以对规划系统给出的结果进行细化处理。指导这项研究的基本理念是我们可以向优秀的案例学习；没有必要去重新开发建筑平面布局，因为现存的类似案例已经是海量的，并且种类丰富。这种理念催生了创建一个拥有大量精心构思的可行性建筑平面的数据库想法，最终演变为在 10 种不同规模的住宅布局间的选择，这其中包括 Patrick Gmür 和

better houses than an architect or an urban planner can: the important aspect is that many solutions can be generated quickly, compared to one another and finally evaluated, whether the criteria are the costs per square meter or the amount of evening sunshine received per square meter. Qualitative criteria like noise exposure or light incidence can, using these procedures, be definitively addressed and quantified.

A defining characteristic of such a large project—and this is where we come to the second phase in our digital chain—is the question of the individuality of the ground plans. These days, individuality is a distinctive human need, especially when talking about private dwellings. Everybody wants an individual, made-to-measure environment. In the case of a large residential development like Hardturm, this would hardly be possible using traditional planning constraints, since individual ground plans could be implemented with difficulty at the building or planning stage, within the limits of a justifiable resource investment.

For the Hardturm study, we developed a procedure that enabled the provision of individual ground plans and which presented the next logical step for the detailing of the results that had so far been obtained from the planning system. The basic idea is that one can learn from good examples; that it is not necessary to reinvent successful ground plans, since they already exist in sufficient numbers and with sufficient variety. This idea led to the concept of a library of well-conceived workable ground plans, and ultimately to a choice of ten dwelling layouts of various sizes, from architects

Gigon/Guyer 在内的多位建筑师的成果。这些平面图现在必须要被适当修改以适应上述区域内的指定建筑地块。为达到这个目标，各种不同的设计标准起着至关重要的控制作用，例如建筑的入口和立面，噪音量和日照度。而后系统在这个可能的平面布局上实验建筑规模、拓扑结构和各类控制性参数，直到找到最佳的形式。第一次调整一般是根据建筑规模做出的。在第二步时，可能的建筑平面布局会被翻转扭动以适应预先设定的规划条件。建筑的窗户和墙体的大小会被不断调整，它们之间的角度也会不断变化，并且建筑的门和室内的家具位置也会调节变动。这个过程最大的优点在于整个数据模型保持着连贯性，所以在实验中任何阶段的整套数据都是可以被选定和实践的，尽管居住区平面图的最终定案是要考虑客户的需求，根据销售来决定的。

然后我们到达第三步，也是最终定案的阶段：立面的构筑。在这一步中，我们可以利用计算机来设计建筑的立面，这种方式不仅可以满足客户对于个性化和独特构思的设计方案的要求，同时可以有效控制生产的费用不超过投资预算。这同时宣告了简单的、无差别的重复式立面时代终结了。就 Hardturm 研究案例来说，立面设计的另一个方面同样十分重要，即整合立面设计的模块化建造可以独立地实现，这与整个项目内在的可变性和程序性理念相一致。这就满足了整个项目的可变性和程序化特征。提供的模块尽量标准化，每个模块服务的生活空间在 15 平方米左右，通过软件可以将它

such as Patrick Gmür or Gigon/Guyer. These ground plans now had to be tailored to fit into the designated building plots of the usage zoning plan described above. To this end, differing criteria played an important role, such as entrance and facade, noise level and insolation. The system then experimented with the size, topology, and defining parameters of the possible ground plan, until the best variant was found. A first adjustment could be made based on size; in a second step, possible ground plans could be turned, flipped, and adjusted to the planning conditions that had been laid down. Windows and walls could be made larger or smaller, the angles between them changed, and doors or furniture moved around. The great advantage in all of this is that the entire data model remains coherent and can be checked and implemented at any point, even though the ground plan of a dwelling is only finalized with its sale, taking into account the requirements of the buyer.

As a final step, we move on to the third phase: the construction of the facade. It is possible, using the computer, to design building facades that fulfill the client's wishes for an individual and unique design, but at the same time can still be produced within budget. The time of simple, undifferentiated, repetitive facades is over. For the Hardturm study, a further aspect of the facade design—consistent with the idea of variability and programmability inherent in the entire project was also important: the modular building services that were integrated into the facade could be supplied individually. This met the variable and programmable character of the

21

拼合的平面图被自动转译为建筑立面上的韵律。建筑结构内部的任何变化都会被直接反映在建筑的墙体和窗洞的关系上。另外,这两种设计版本反映出不同的立面程序控制系统能够覆盖相同的建筑结构,在某种意义上与文本可以以很多不同的方式打印出来是同样的道理。与建筑的平面生成的原理相同,建筑的立面系统也来自于一个数据目录。

The patchwork of ground plans was automatically translated to the rhythm of the facade. Any changes in the internal structure of the building would be reflected directly in the changed relationship between the wall and opening in the facade. Furthermore, the two versions show that differently programmed facade systems can clad the same building structure, in a similar manner to the way that a text can be printed in many different ways. Like the ground plans before them, the facade systems were chosen from a digital catalog.

们各自调整修改以适应它们所要服务的空间。根据这个理念，建筑所包含的复杂性被整合到预先设定好的立面单元和服务模块中，建造过程的复杂度大大降低，同时潜在错误的发生率也降低了。与此同时，建造的速度和质量大大提高，建筑的适应性和功能性也得到很好的改善。立面之后的空间使用功能的变化相对于我们的技术来说已经不算是问题，因为通过相关软件就可以轻松调整改变。

这样，整个数字化周期就完成了。研究得出的成果是一个灵活可变的设计，无论是城市规划层面还是最小的建造细节，我们都可以通过调节按钮来改变浴室门的位置并关联到整幢建筑——建筑的整体及其细节可以同时处理。在这个处理过程中，如果说我们应该吸取一些教训，那就是设计并非总是需要小心的灵机一动的创意——考虑到设计人员所做出的努力，这种观点或许还是可以被理解的。如今设计可以被理解为一种可以随意变更的、动态的过程，一方面可以给予设计人员更大的自由度，另一方面，这也意味着承包商和投资商的财务风险大大降低了。

entire project. The supply modules were made as standardized as possible, and each served around 15m² of living space and could be individually tailored through software to the spaces that they served. Thanks to this idea, which incorporated most of the complexity of the building into the ready-prepared facade elements and service modules, the complexity of the building process was significantly reduced, as was the possibility for the introduction of potential errors. At the same time, the quality and the speed of construction was increased, as was the flexibility and the functionality of the building. Changes in the use of the spaces behind the facade presented no great problem for the technology, since it could be easily reprogrammed via the software.

So, the digital cycle is complete. A flexible design is created, from the urban planning level down to the construction of the smallest details, simply because now we can change the position of the bathroom door in relation to the whole building at the flick of a switch—the details as well as the whole project can be dealt with simultaneously. If there is a lesson that we can learn from this, it is that design does not always have to be a one-off ingenious act of creation that must be jealously guarded—a view that is, perhaps, understandable in light of the great effort expended. Design can now also be understood as an open-ended, dynamic process, which, on the one hand, can provide more freedom for the designer and, on the other, can mean fewer financial risks for the contractors and investors.

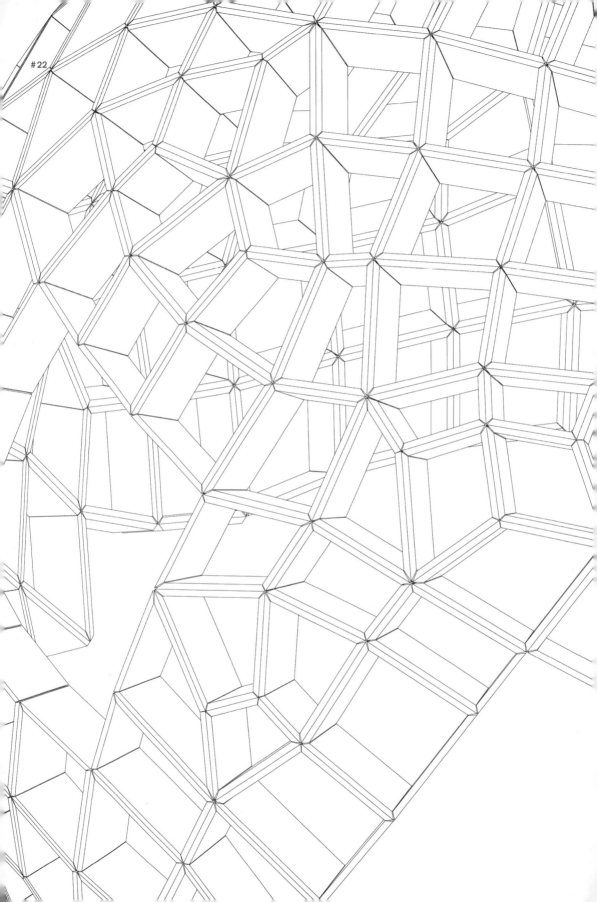

第二章

超越网格

　　网格始终是建筑中的重要工具。它帮助确立一种宏观的视角，并将建筑形成过程中的无数影响因素有序结合。因为已经确立的一些在成本、时间和质量上的优势，它已然成为建筑工业化生产时代的基础。模式和装饰同样满足这个功能。几乎所有的网格和模式都有这样一个特点，即它们可以被计算出来，因此它们是可预计并且是明确的：利用最少的几种不同的元素就可以创造出可控和紧凑的表面或体量，或者是通过增加一些个性化的元素例如砖来创建。这种应用比较简单，这意味着运作的可预测性和可计算性，举例来说，运作可以是基于两个自然数之间的关系，例如1:2、1:3等。有了这种关系，表面可以被逻辑化地建造而成一定比例。简单的网格系统的另一个特点是通过整体控制细节。这就造成一种典型的局面，即在网格系统内不允许出现相对坐标的定义。任何一个特定的坐标元素一定无法区别于其他任何坐标元素——尤其是当人身处于这种网格之中。

　　在这个问题上值得一提的是1974年英国数学家Roger Penrose开发出的第一种与众不同的铺面，一种不规则的、无法预期的铺排形式，但是依然可以完全覆盖整个表面：这就是所谓的Penrose铺面。这个系统仅仅包括两种不同的面砖，但这两种面砖的组合却可以形成规则、可预期的铺面方式不可能实现的立面造型效果。仅仅是通过排布这些面砖，就能展现多样的排列组合的可能性，产生的形式是一种不可预知的、无理性的网格系统。在Penrose向人们展示这一成果之前，这一现象一直是完全不为人知的，令人惊奇的是发现这一现象竟然花费了这么长的时间。

　　现在让我们回到传统网格系统的讨论。在20世纪中期，勒·柯布西耶在他的《模度》中创造出一种建筑图表，他按照黄金分割比例划分人体的尺度比例，并将它们转化为一种理性的关系。在误差幅度百分之一的范围内，勒·柯布西耶估算出这个比例为5:8，虽然这个数据不如1:2容易控制，但相对于黄金分割这个难以控制的无理数来说，这个控制性比例还是相当容易实现的。

　　虽然我们已经远离了预制楼板时代，网格系统在今天的建筑设计领域依然发挥着非常重要的作用。如今，网格被旋转和变形，创造出杰出的建筑，并同时保持着组织上和建造上的控制性。弗兰克·盖里在毕尔巴鄂的古根海姆博物馆，或是诺曼福斯特在伦敦的瑞士再保险公司大楼都是利用网格变形设计的杰出案例，在这些设计中，网格的边缘被放大、缩小或是旋转。在这个案例中我们可以观察到一种范式的变革，这一变革被当代信息技术所驱动：网格系统不断调整其自身以适应形式，而不是形式去适应网格系统，后者我们可以在拙劣的预制楼板或是孩子的拼装玩具中看到。网格系统的适应性不断加强，它原先的功能性原则颠倒了，但并没有根本性改变。这种翻转的网格系统几乎是我们现今所能做到的极限，它标志着一条分界线，网格系统中更大的自由度难以达到。它同时也标志着建筑生产走出了工业时代，踏进信息时代。

　　在信息技术的辅助下，我们现在已经可以推翻网格系统在建筑领域的统治性局面。如今，已经不单单是网格的边缘可以变化：网格元素本身也获得了更大的自由度，它们可以挣脱网格进行自由组织。规模、比例或是它们邻近单元的数量都可以不断地重新调整，而不丧失整个系统的完整性。

　　最大的区别在于运转模式的不同。对比网格系统，信息技术在建筑设计中的作用不仅仅在于作为等级分明的一维组织原则，它还使得相互依赖和互相平衡的网络和结构具备技术可识别性。信息技术并非根据我们在力学中得知的那些已经被确立的、严格的因果关系来发挥作用的，而是根据信息交流和相互的动态交换规则来运行的。

　　在俄国数学家 Georgi Voronoi 的图表中，我们发现一种十分明显的平面组织原则，它符合信息技术所呈现出的可能性。图表没有以网格系统的形式表达（尽管图表可以创造出网格系统），它们的特征更类似于泡沫的结构。各个独立的元素相互沟通协调、胀大或是收缩，改变自身位置或是从一个位置消失然后重新出现在别处。这已经不单是网格边缘的连续性或非连续性变化的问题，同样也关系到了变化的邻里单位。系统的每个单元都在变化，因为无论这个变化的单元是多么微小，它都要求整个系统的重新构建，所以这种类型的系统难以手绘记录，甚至传统的 CAD 系统也无能为力。另一方面，正确编程的计算机，可以保证这些灵活、动态的区域系统处于平衡状态。超越网格系统，建筑的易变性和可塑性都大大加强。或者说，正如我们在 CAAD 实验室所热衷于阐述的：经典的网格系统是凝固不变的；Voronoi 图表是自由可变的。

Beyond the Grid

The grid has always been a central device in architecture. It helps to maintain an overview and to keep order while combining the countless elements from which every building is made. It serves as the basis for the industrial production of architecture with its established advantages in cost, time, and quality. Pattern and ornament also fulfill this function. A feature of almost any grid, and any pattern, is that it can be calculated, hence its predictability and clarity: with a minimal number of different elements it can be guaranteed that a surface or volume can be created in a controlled and tightly packed fashion, or it can be built by adding individual elements such as bricks. It is a trivial, meaning a predictable and calculable operation, which works, for example, on the basis of the relationship of two natural numbers, with which relationship surfaces can be logically structured and proportioned, as in 1:2, 1:3, and so on. A further characteristic of simple grid systems is the fact that the totality defines the details. This leads to their typical appearance which, within the grid, does not allow the definition of relative positions. One particular grid element cannot be distinguished from any of the others—particularly when one is sitting or standing inside the grid.

It is worth mentioning at this point that in 1974 the British mathematician Roger Penrose developed the first non-trivial tiling: in other words, an irregular, unpredictable arrangement, but one which nevertheless completely covered the surface: the so-called Penrose tiling. The system consisted of only two different tiles, but ones that are shaped in such a way that a regular, predictable tiling of a surface is simply impossible. Only by actually laying out the tiles, a process which presents ever more possibilities for their ordering, does the real pattern emerge—an unpredictable, irrational grid. Before Penrose formulated it, this phenomenon was completely unknown and it is surprising that discovering it has taken so long.

Let us now return to the conventional grid. In the middle of the last century, Le Corbusier created an icon of architecture with his Modulor, in which he divided human measurements and proportions along the lines of the Golden Section, and converted them into a rational relationship. With a margin of error of about one percent, Le Corbusier reckoned this ratio to be 5:8, which, while it is not as easy to deal with as the ratio 1:2, is considerably easier than the unmanageable Golden Section, an irrational number.

Grid systems still play a considerable role in architecture today, even if we have long left behind the days of Plattenbau. Today, grids are rotated and deformed, creating spectacular architecture, while still retaining organizational and constructional control. Frank Gehry's Guggenheim Museum in Bilbao, or Norman Foster's Swiss Re Tower in London are prominent examples of the deformation of a grid, in which the edges are made larger, smaller, or rotated. In this we can see a paradigm shift, one driven by modern information technology: the system, the grid, adapts itself to the form, rather than the form to the grid—something that we see in the infamous Plattenbau, or in our children's Lego sets. The system becomes more adaptive, its previous functional principle turned upside down, but not fundamentally altered. This turning-around of the grid system is about as far as we can go, and it marks a dividing line, since more freedom is simply not available in a grid system. However, it also marks architectural production's departure from the industrial age and its entry into the information age.

With the help of information technology, it is now possible to overthrow the tyranny of the grid. Now, it is not only the edges of the grid elements that can vary; the elements themselves gain considerably more freedom. They can break free of the grid and organize themselves freely. Their size, their proportion, or the number of their neighbors can be continually renegotiated without losing the integrity of the overall system.

The biggest difference can be seen in the mode of operation. In contrast to the grid, in architecture, information technology not only functions as a hierarchical, one-dimensional organizing principle, it also allows for the technical visualization of networks and structures that are mutually dependent and held in equilibrium. Information technology does not function according to the established, strict process of cause-and-effect that we learn from mechanics, but rather according to the rules of communication and of mutual dynamic exchange.

In the diagrams of the Russian mathematician Georgi Voronoi, we find an obvious organizational principle for surfaces that corresponds to the possibilities presented by information technology. The diagrams do not behave like grid systems (although grid systems can be created using them), their behavior is more akin to the structure of foam. The individual elements communicate among each other, they grow or shrink, they change their position, or they disappear from one location, only to reappear in another. It is no longer a question of the continuous or discontinuous deformation of the edges of a grid, but also one of changing neighborhoods. These types of systems cannot be drawn by hand or even with conventional CAD systems, since every change, no matter how small, would require that the entire system be newly constructed. Properly programmed computers, on the other hand, can keep these flexible, dynamic zoning systems in balance. Beyond the grid, architecture becomes more fluid, more ductile. Or, as we in the CAAD department like to put it: classical grid systems are clotted; Voronoi diagrams are creamy.

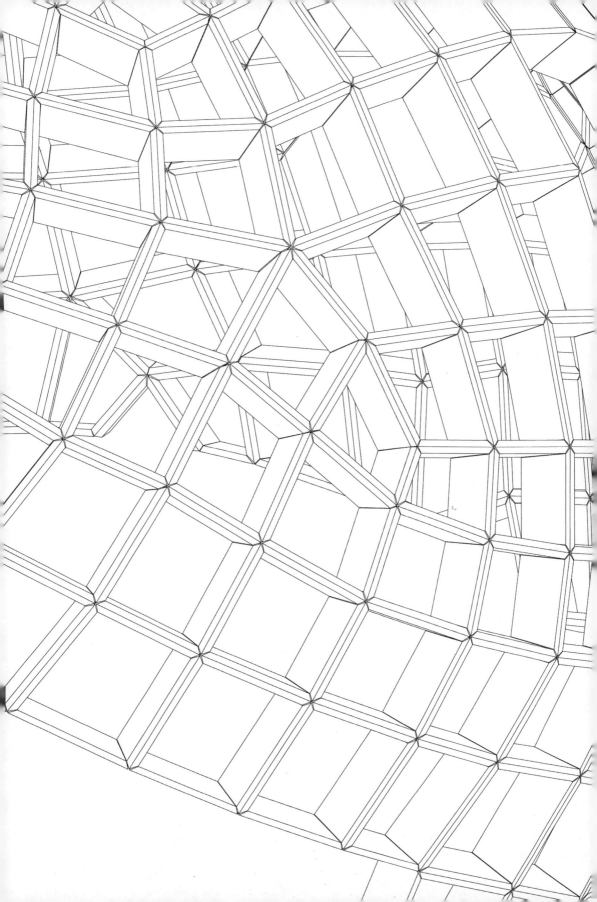

Michel Serres:
Hermes III

' They have set off from all the old places on the map of the world.
Where there are cities, where there are islands, cases, boxes, black holes, niches,
cores, groups, fields, milky ways, and pinheads.
Chains of galaxies and spheres of almost unending gravity.
Parabolas whose major axis is as gigantic
as their minor axis is small for those lakes without horizon.
They had always known it, but never knew it and experienced it now:
that space is unexpected.
Populated, teeming, foreign, and wondrous.
Everywhere impossible, heterotopic. Space is like Pandora's box.
Numerable accumulations of heterolitic beings,
from the fits of the trivial to the unimaginable,
from the black-body radiator to the ensnarement of the elements,
from the Euclidean orbit to the pathological discrepancy.
Everything is made new by the suns. '

模式

Patterns

项 目： **The Millipede**

时 间： 2002

指导老师： Markus Braach

学 生： Oliver Königs

拐角 Around the Bend

拐角是一个我们经常探讨的问题。它的处理通常比较困难，例如，如果不是切开的话，拐角处的墙纸就会破坏图案的规律性。在维也纳的维特根斯坦住宅就是一个最终失败的尝试。它试图在从内到外的各个细节层次上有规律地组织尺度。只要是采取像墙纸那样零碎的、有规律的、可预测的模式，任务就注定要失败。

通过千足虫程序，我们可以使这一问题的解决变得轻而易举。这一窍门是摒弃墙纸图案原有的规律性样式，而是以一个反馈过程开始。乍看之下，这种小型研究项目对于熟练的程序员是一个简单的任务：一种小型自动装置，装配有用户规定数量的传感器，在既定平面上移动。它移动的规律在原理上非常简单：已经过的区域会影响它下一步的方向，而且不能超出表面边界。用奥地利物理学家、自动化专家 Heinz von Foerster 的话说，千年虫就是我们理解这个"非凡机器"的生动例子。千足虫在平面爬行形成的轨迹而得到的图样是个运动产物，因而不能被预测，因为只有经过的区域才能决定下步是什么，除非它在原有起点状况下重新出发。可是，如果千年虫从一个不同的位置出发，就会形

Corners are always a problem. It is often difficult, for example, to wallpaper around corners without having to make cuts and therefore destroying the regularity of the pattern. The Wittgenstein house in Vienna is the showpiece for an ultimately unsuccessful attempt to regularly organize dimensions, at all levels of detail, inside as well as out— a task doomed to failure, as long as one is working with trivial, regular, and predictable patterns like those found in wallpaper.

With our millipede, we can solve this problem with relative ease. The trick is to abandon the regularity of the wallpaper pattern and instead to start with a feedback process. At first sight, this little research project would appear to be a simple task for an experienced computer programmer: a type of small automaton, equipped with a user-definable number of sensors, moves over a predefined surface. The rules to which it must adhere, are, in principle, very simple: stretches that have already been covered will influence the direction that the automaton will take, and the boundaries of the surface may not be crossed. And so our millipede is a vivid example of what we— in reference to the Austrian physicist and cyberneticist Heinz von Foerster—understand as a 'nontrivial machine.' The pattern that

The Millipede

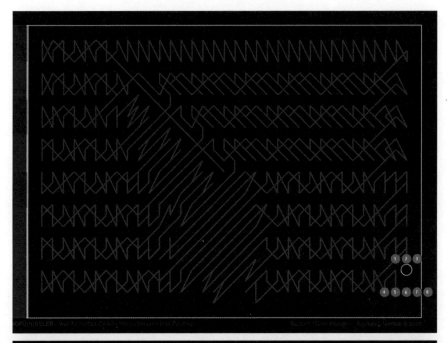

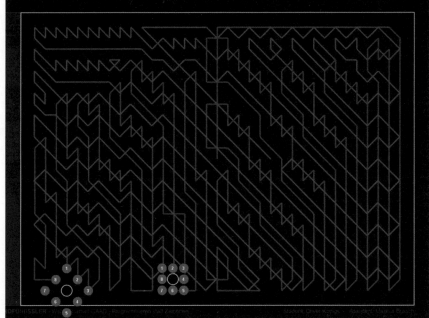

千年虫是我们给自动装置的命名（图中用白圈表示），它在任意表面爬行，经过处形成一条路线。通过传感器（用小红圈表示）观察其周遭环境，确保自己绕开阻挡物。当它已走过的轨迹被程序设定为障碍时，它的轨迹时就会形成新的、无法预测的模式，不同于预先的安排。

The millipede is a name we gave to an automaton(represented in the picture by a white circle) that'crawls' across an arbitrary surface, drawing a line as it goes. With its sensors (shown as little red circles), it 'sees' its environment, enabling it to avoid obstacles. When the millipede is programmed to interpret its own trail as an obstacle, new and unpredictable patterns emerge, independent from the arrangement.

54 模式 Pattens

#25

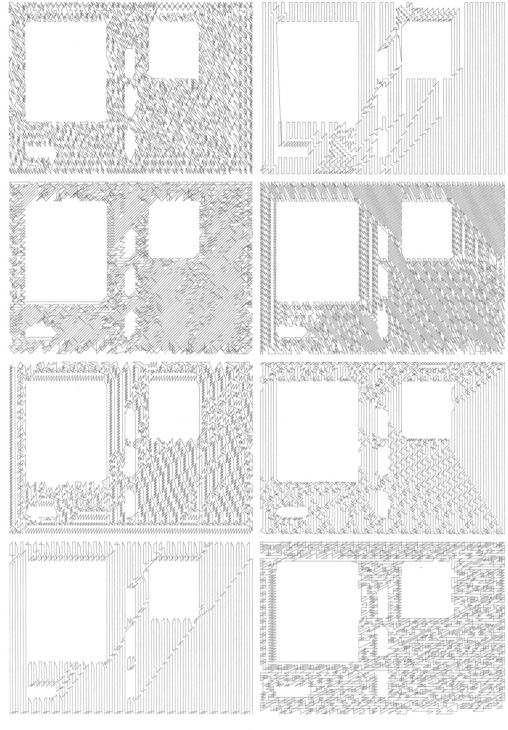

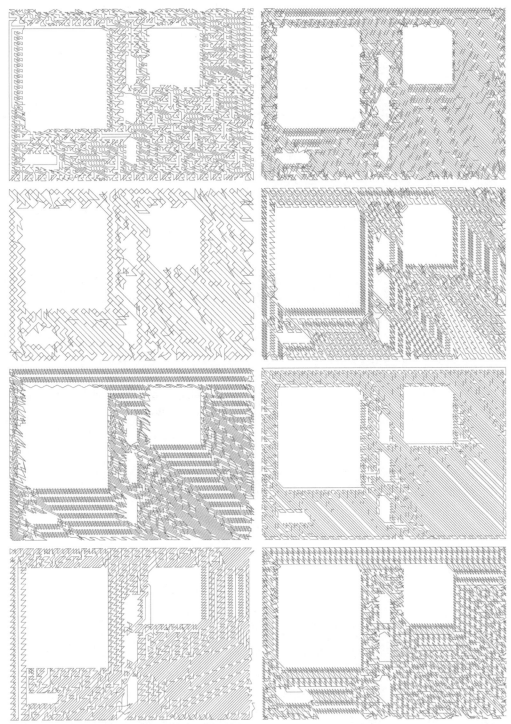

运用数字技术，建筑设计元素，尤其是建筑装饰，开启了新的美学维度。这让人联想到自然系统：表面上的例外与常规之间的区别消失了。

Using digital technologies, architectural design elements—especially ornamentation—take on a new aesthetic dimension, reminiscent of natural systems: the difference between exception and regularity disappears across the surface.

成另一种运行模式，这取决于传感器的位置和数量以及平面的几何形态。它下一步的实际运行轨迹可能无法预测，但它一定能达到我们想要的目的。这种简单而特别的系统特性就是它坚定不移地执行自己的任务，因此那些障碍物、拐角以及边缘都不会阻碍它的移动。这种方式产生非常错综复杂的模式，一方面是重复与规则的结果，另一方面也是遇到障碍物机动规避的产物。

从技术角度来说，千年虫程序包含了多个分级编程、自由配置的传感器，是一个简单智能系统。如果第一个传感器的路径被阻隔，那第二个传感器的路径会被采用，以此类推。然而，一旦第一个传感器重新找到可行路径，这条路径就会被采用。千年虫编程的特点是，它不是多重的、互动的智能体，系统内只有一个活跃的智能体与不断变化的环境进行对话，但同时它只能了解它最近的环境。这种意识范围的局限性就需要程序员来弥补。然而这种特性确实与实际生活相符合，无论作为标准的是人类视野、工业机器人的操作区域，或是群体系统。在这些情况中，单体在适应其紧邻的邻居时起作用，而无需将系统作为一个整体。

the millipede traces out on its journey over the surface cannot be predicted and is a product solely of the millipede's movement, since only the stretches already covered can determine what the next step will be—unless one starts off again with exactly the same starting conditions. However, if we start the automaton from another position, a different pattern of movement will be generated, dependent on both the position and number of sensors and the geometry of the surface. The actual path followed may be unpredictable, but the arrival at the goal is not. A characteristic of this simple, but still nontrivial, system is that it rigidly sticks to its task, so that obstacles, corners, and edges do not impede its journey. In this manner arise delightfully intricate patterns, which are a result of repetition and regularity on the one hand, and obstacles and evasive maneuvering on the other.

From a technical point of view, the millipede is a simple agent system with hierarchically programmed, freely configurable sensors. If the way covered by the first sensor is blocked, then the way covered by the second sensor is taken, and so on. However, as soon as the first sensor again detects a free passage, then that way will be taken. Characteristic for the programming of the millipede is the fact that, instead of multiple, interacting agents, there is only one active agent in the system that communicates with a continually changing environment, but which is at the same time aware only of its immediate surroundings. This limitation on its scope of awareness actually places great demands on the programmers. It does, however, correspond to reality, whether we take as a criterion the human field of view, the operational zone of an industrial robot, or even flocking systems. These also function only when their individual elements orient themselves directly to their immediate neighbors, without taking into account the system as a whole.

像千足虫这样的系统非常有趣，因为系统内部很简单，只是传感器和运动之间的简单而可预知的组合。如果从外部来看，效果和最终的装饰异常丰富而不可预测。这种特性为系统在建筑中的日常应用打开了有趣的视角。直到现在，建筑装饰的特征都是相同模式的不断重复——这样，我们在处理拐角、边界和对称时就会带来问题。千年虫程序可以处理所有这些问题，因为它可以与环境进行沟通，并用一种连贯和逻辑方式作出反应。

A system like the millipede is all the more interesting, since the internal view of the system is trivial and can be detected via a simple and predictable combination of sensors and movements. However, when seen from the outside, the effect and the resulting ornaments are anything but trivial and predictable. This characteristic opens up interesting perspectives for the system's everyday application in architecture. Until now, ornaments were characterized through the repetition of the same pattern—something which leads to problems when we have to deal with corners, edges, and symmetries. A programmed millipede can deal with all these problems, since it can communicate with its environment and can react to it in a consistent and logical manner.

项　目：**Semper Rusticizer**

时　间：2003

参与者：Rusell Loveridge, Kai Strehlke

追踪 Chased Out

建筑风格在不同尺度上会给人不同的感知，该项目便是个很好的例子。从远处看，石棉水泥板给人"粗糙石面"的感觉，与 Gottfried Semper 原始的凿刻石面没有什么不同。只有在近距离观察时，才能发现它的表面纹理是通过数字技术制造出来的。我们可以进行深一步探讨：不单单是纹理和表面处理可以被程序化，复制品和原件在多远距离上能被区分开也可以通过程序来控制。

Architecture is perceived differently at different scales, and this project is a particularly fine example of that. When viewed from afar, Eternit panels that have been given a 'rusticating' finish, look little different from Gottfried Semper's original chisel-carved stone rustication. It is only when one draws nearer that the surface texture created by the digital rustication process becomes apparent. But we can go further: It is not only the texture and the treatment of the surface that can be programmed, but also the distance at which the reproduction and the original become distinguishable.

这一项目起始于苏黎世瑞士联邦理工学院主楼北外墙 Gottfried Semper 设计的粗面砌筑。它的目标是利用现代的、计算机控制的设计和生产工具重塑这种砌面方式。从外墙细部的高清数码照片出发，通过灰度分析，可以推断石头的颗粒度，同时，通过分析它的投影，可以了解表面的拓扑结构。此外，程序还遵循了用凿子加工石头的模式。这些颗粒度、拓扑结构和工艺上的数据用来生成各种各样算法，创建铣削加工的轮廓。我们用石棉水泥板取代了 Semper 时代的石头（当时没有任何结构功能）。就像我们所说的，这种制作过程的动人之处就在于，如果从远处看，原件与复制品几乎无法区分，而区分两者的实际距离

The starting point of this project was Gottfried Semper's rustications on the north facade of the ETH main building in Zurich. The aim was to recreate this rustication using modern, computer-controlled design and production tools. High-resolution digital photos of sections of the facade served as the starting point of this project. By analyzing the grey values of the images, it was possible to deduce granularity of stone and also, by analyzing the way that it cast shadows, to get an idea of the topology of the surface. Furthermore, programs were also able to follow the pattern of the working of the stone with a chisel. The resulting data on granularity, topology, and workmanship was used to drive various algorithms used in the

由 Gottfried Semper 设计的瑞士苏黎世联邦理工大学主楼，有着粗糙石面的基座。

The main building of the ETH by Gottfried Semper, with its rusticated plinth.

粗面石板的高清数码图片。这些图片是我们这个项目的起点。

High-resolution digital photograph of a rusticated stone block. Such photographs were the starting point for our project.

#28

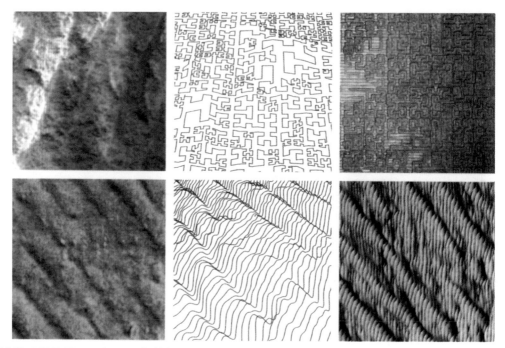

高清数码图片显示了粗面石板的拓扑结构、颗粒度和纹理。算法将这些数据转换为 CAD 系统中的代码，再反过来控制铣削／切割机的切头，产生石棉水泥板版本的粗面砌筑复制品。

High-resolution digital photographs show the topology, grain, and texture of a rusticated stone block. Algorithms turned this data into a representation within a CAD system. This in turn controlled the cutting head of a milling / routing machine, which produced a 'reproduction' of the rustication in Eternit.

#29

相较于复制，比如模铸，算法的变革代表着计算机带来了不寻常的可能性。最终的设计结果可根据目标而灵活多变。

In contrast to a reproduction—a cast, for example—the transformation into the world of algorithms offered the possibility of computable alienation and the design of an end result that was as targeted as it was flexible.

则可以通过程序来简单控制。如果需要，复制品也可以做得与原件极其相似。

该项目可以体现出传统手工过程与计算机控制过程之间的多重差异。当然，那种用传统工具（锤子和凿子）、传统材料（石头）和传统方式（手工）生产的粗糙石面外墙的工作形式，从经济角度上说，已经不再可行。但如果我们用现代工具如电脑控制的铣削机械和切割工具，就可以产生独特定制的表皮和形式，并且控制在成本预算之内。但问题也随之而来，用现代机器来模仿古代生产方式总有点不大协调。像"千年虫"和"粗糙石面"这样的项目告诉我们，因为有了数字技术，在建筑装饰领域有了更多新选择。

creation of milling profiles. Instead of using stone, as was the case in Semper's time (and which did not have any structural function in the original), we substituted Eternit panels. As we have already said, the beauty of this production process is that from a distance the original and the reproduction can not be distinguished from each other, that the actual distance at which the difference becomes noticeable is dependent simply on programming, and that the original can be reproduced extremely accurately, if desired.

In a project like this, it is possible to experience the multi-dimensional difference between traditional, hand-worked procedures and their implementation through a computer-controlled process. Of course, producing surface working like that found in rusticated facades using traditional tools (a hammer and chisel), the traditional material (stone), and the traditional method (by hand) is simply no longer economically feasible. However, if we employ modern tools like computer-controlled milling machines and routers, individual, bespoke surfaces and forms can be produced within the budget of industrial production. However, the problem arises that it does feel somewhat dissonant when old working practices are imitated using modern machines. Projects like the millipede and the 'Rusticizer' show that, thanks to digital technology, a broad range of new considerations is emerging in the field of ornamentation.

项　目：**Processing**

时　间：始于 2004

指导老师：Christoph Wartmann

学　生：Marc Zürcher, Rahel Metzger, Martin Staubli, Sarah Eva Brichler, Karin Gauch, Fabien Schwartz, Andy Plüss, Martin Jakl

编程而非绘图 Programming, Not Drafting

我们可否让建筑师开始学习程序语言？当然，让建筑师为了建筑成为真正的软件程序员并没有什么意义。但拥有程序语言能力可以让我们熟悉另外一种思维方式，快速认识到周围环境中的技术性本质。

Can we expect architects to start learning programming languages? Of course, asking architects to become true software programmers for architecture would not make much sense. However, an ability with programming languages is a good way to gain familiarity with specific methods of thinking that increasingly reflect the technological reality of our surroundings.

给予建筑师一种能让他们尽快熟悉起来的语言对于我们来说很重要。"Processing"软件出现以后，我们认为我们找到了这样的工具。它学起来很快，仅仅一小会儿就能取得成效。Processing 从不忽视基础理念：基于 Java 的程序语言，由波士顿麻省理工学院的 Ben Fry 开发，开发环境非常简单，这意味着不论是初学者还是有抱负的程序员都能快速掌握它。

It is important for us to be able to offer architects a language with which they can quickly become familiar. With 'Processing,' we believe that we have found such a tool. The learning curve is, for the most part, very steep, but after only a very short while, successful results are being produced. Processing does not trivialize the fundamental concepts: a Java-based programming language, Processing was developed by Ben Fry at MIT in Boston, and presents a very simple development environment, indicating that Processing has enormous potential for both beginners and more ambitious programmers alike.

一些应用领域的例子就足以说明问题：Processing 可以改变大量图像和音频参数，甚至可以开发乐器。通过一些指令，可以播放 mp3 文件或是电影，图像也可被加载和操作。Processing 可以

A few examples of the range of possible uses should suffice: Processing can change various image and sound parameters and even develop musical instruments. With a few commands, MP3 files or films can be

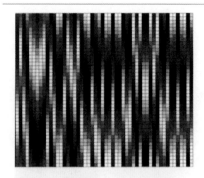

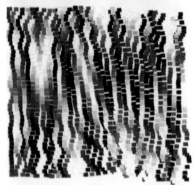

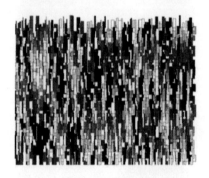

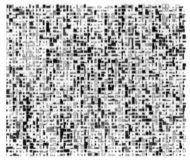

2004 年由 Michelangelo Ribaudo 完成的学生作品。
左侧是一系列算法生成的图像；右侧是 Processing 计算机语言相应的算法。由算法生成图形，源代码中很小的变化也会对最终的图形结果产生相当大的影响。这是一个很有意思的特征。

A student project by Michelangelo Ribaudo, 2004.
On the left is a sequence of algorithmically generated images; on the right, the corresponding algorithms in the Processing computer language. As is often the case with algorithmically generated graphics, even small changes in the source code produce considerable effects on the graphic results—an interesting characteristic.

#31

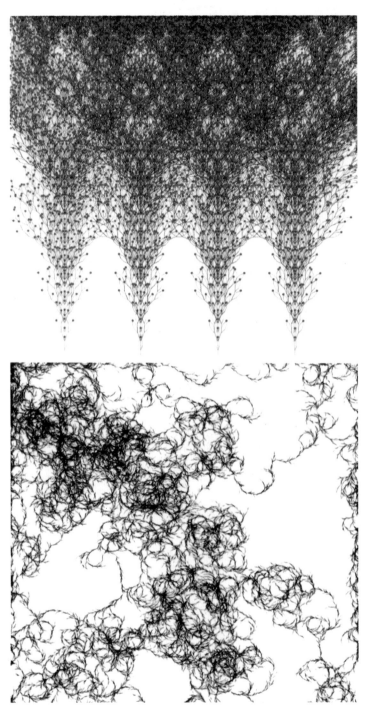

2004年由Michael Idoine 完成的学生作品。几行源代码形成了典型的、看上去有机的"Lindenmeyer system"。它可以用来计算植物生长以及其他可以被递归定义的生长过程。通过少量参数的变更，同样的源代码可以形成明显区别的图像。

A student project by Michael Idoine, 2004. A few lines of source code generate the typical, organic-looking 'Lindenmeyer systems.' These can be used to calculate plant growth and other recursively defined growth processes. After altering only a few parameters, the same source code can produce a distinctly different image.

制作快速变幻的图形，或创建人可以穿行其中的3D场景。此外，Processing还可以与互联网上的服务器或数据库建立连接，甚至建立自己的网络。通过Processing，你可以很方便地控制一系列的用户互动装置。我们还和学生一起创建输入和输出装置，最终会实现无线操作。

像面向过程的设计或面向对象的设计这样的编程概念，结合智能体系统，可以用Processing简单快速地验证。John Horton Conway的生命游戏，一个经典的细胞自动机的程序，可以用Processing快速而简单地运行。这个二维的游戏界面由一组细胞组成，细胞的状态可被定义为生或死。游戏的规则非常简单：当一个细胞周围有2或3个细胞相邻时，该细胞将存活；当细胞周围的细胞数量低于2或大于3，该细胞就会死亡；当死细胞周围恰好有3个细胞相邻时，该细胞将在下一代获得重生。这三个简单规则就可以形成一个复杂、不可预测的互动模式。蚂蚁或鸟群行为系统，也一样可以被很轻易地模仿出来，宏观结构的形成，即所谓的涌现性，就可以被生动地展示出来了。

played, or images can be loaded and manipulated. Processing can generate fast-moving graphics or 3-D that can be walked through, again using only a few commands. Furthermore, Processing can establish connections to servers and databases on the Internet, or link up its own networks. With Processing, you can quickly control a large array of devices for user interaction. We are also building input and output devices with our students that will, in time, be operated wirelessly.

Programming concepts like procedural and object-oriented programming, combined with agent systems, can be quickly and simply tested using Processing. John Horton Conway's Game of Life, a classic among the cellular automata programs, can be implemented quickly and easily using Processing. The two-dimensional playing field is composed of a field of cells that can either be defined as dead or alive. The game operates using only a few short rules: Any cell with two or three neighbors is alive; any cell with fewer than two or more than three neighbors dies; any dead cell with exactly three neighbors will come back to life in the next generation. These three simple rules are sufficient to generate a very complex, unpredictable interplay. Systems that simulate the behavior of ants or the flocking of birds can be simulated just as easily, and the formation of meta-structures—the so-called emergent properties—can be vividly demonstrated.

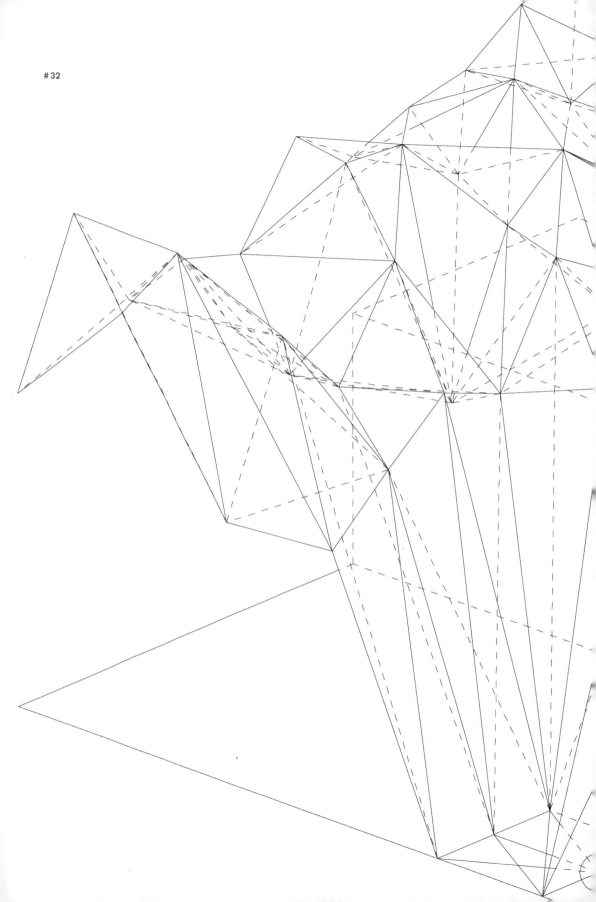

第三章

不可预估的设计

上一章列举的案例展示了我们这种方法的一个基本特征：电脑辅助设计使我们放弃了对最终结果的控制，这意味着在最后结果出来之前我们不知道建筑是什么样子，甚至不知道它为什么是这个样子。计算过程中经历了太多不同形式，我们无法追踪它的全部。但我们知道最终产生的结果取决于我们最初编写的程序规则。这种思维方式的转变对建筑师来说可能需要一定的时间适应，所以我们要先把背景介绍一下。

分级组织系统或线性因果关系主导的系统除外，我们可以运用的规则很多。正如我们开始处理网络，或将其发展成为以目的为导向的方式，那么我们通常用来处理机械系统的规则就完全不适用了。网络交互系统中的一个非常有趣的现象就是涌现。实际上人们很早就观察到了这一现象。整体不仅是各部分之和，因为各部分的动态交互性带来了一种新型结构或行为，它们不能由最简单的规则来解释。涌现系统，最本质的特征是创造性。当从外部观察时，涌现性是显而易见的；而从内部观察时，只能看见最简单的规则。

通过使用那些能够遵循特殊路径和躲避障碍物的简单、原始的机器人，可以看出在自然界广泛分布的涌现现象也适用于技术应用。Maris 和 te Boekhorst 发明了一种叫"Didabots"的微型机器人，它配备有小马达驱动的两个轮子和两个可以用来避开障碍物的传感器。这种微型机器人的起源可以追溯到 Valentin Braitenberg 演示的简单车载控制实验。Braitenberg 的车载实验最先演示了看上去复杂的行为并不需要复杂的规则来控制，而这种微型机器人则显示了突发性的、集体性的行为模式。编写一个智能机器人非常简单，当传感器侦测到右边有障碍物时，机器人就向左边移动；没有侦测到障碍物，就保持直线运动。但有趣的是，如果遇到处于两个传感器之间一个过小的障碍物，智能机器人就发现不了，那这个障碍物就会被移动——直到出现另外一个障碍物。这就是从内部发现的系统属性，那从外部看呢？智能机器人把障碍物清理成一堆一堆的。这种未预期的（也不能被预测）行为是涌现产生的。但有趣的是，这种系统行为也是相当稳定的。它不取决于障碍物的数量、是否被移走或新障碍是否出现；它也不取决于在平面上移动的智能机器人的数量（它们将彼此认定为障碍物）或是否有机器人坏掉。此外，涌现现象也不取决于智能机器人所运动的平面是否被置换。在这种不断变化的条件下，相同的涌现事件产生了：机器人将障碍物清理了。

建筑师们有时会害怕这种具有独立涌现性的电脑辅助系统会阻碍个人的创造力。我们觉得这种担心是毫无根据的。诚然，使用这种系统是一个全新的挑战，因为细节行为无法提前预知。可是设计的突发性和结果的不可预知性是激动人心的，尤其是处理那些一般工作方法无法完成的任务时更是如此。我们面临的挑战是进行更深层次、更抽象的思考。未来我们要开发和设置的正是程序，用以控制过程、产生最终形式。

The Design of the Unforeseeable

The examples in the previous chapter show a fundamental characteristic of our approach: our computer-aided work allows to relinquish control over the end result. This means that we do not know what the building will look like until the end of the calculation—or even know why it looks like it does. The calculation process runs through so many variants that we cannot hope to keep track of them all. However, we do know that the results that they produce correspond to the rules by which we initially programmed. This shift in mindset for architects could take some getting used to, so first a few words on the background.

Outside of hierarchically organized systems, or systems organized along a linear cause-and-effect principle, drastically different rules apply. As one begins working with networks, or tries to develop them in a goal-oriented fashion, all those practices that were learned when dealing with mechanical systems are no longer of any use. One of the most exciting phenomena of networked, interactive systems is that of emergence. In principle, the phenomenon has been known about for a long time. The whole is more than the sum of its parts, since the dynamic interaction of the parts brings forth a structure or an event that would never have been suggested by an inspection of the simple rules. Emergent systems are, in the truest sense of the word, inventive. When seen from the outside, the emergent properties are evident; from the inside, however, only simple rules can be seen.

By using simple, primitive robots that can do little more than follow a particular path and avoid obstacles, this phenomenon—widespread in nature—can be shown to also hold true for technical applications. Maris & te Boekhorst have developed so-called 'Didabots,' tiny robots, each having a small motor driving their two wheels, and two sensors that allow them to avoid obstacles. The Didabots can trace their lineage back to an experiment on simple cybernetic vehicles, performed by Valentin Braitenberg. While Braitenberg's vehicles first demonstrated that complex-seeming behavior does not need complex rules, the Didabots show emergent, collective behavior. Programming a Didabot is simple: When its sensor registers an obstruction on the right, the Didabot moves to the left; if no obstruction is registered, it moves straight ahead. Things get more interesting, however, if an obstruction is so small that it lies between the two sensors, thus remaining undetected and getting displaced—until another obstruction crops up. So much for the system properties viewed from inside. But what happens when viewed from the outside? The Didabots clean up. This unanticipated behavior—one that could not have been predicted—is emergent. Interestingly, this system behavior is very stable. It is not de-pendent upon the number of obstructions, whether obstructions are removed or new ones are introduced. It is also not dependent on the number of Didabots moving around on the surface (they identify each other as obstructions) or whether one of them has broken down. In addition, the emergent phenomenon is not dependent upon the displacement of the surface on which the Didabots are moving. Under widely varying conditions, the same emergent effect is produced: the Didabots clean up.

Occasionally among architects the fear will arise that individual creativity will be hindered when working with computer-aided systems that display a certain emergent independence. We feel that this fear is unfounded. Of course, working with these sorts of systems is a whole new challenge in itself, since their detailed behavior cannot be predicted in advance. But it is very exciting to design emergence, to try to channel the unforeseeable, especially when working with tasks that are not amenable to normal working methods. The challenge consists of going one step further in working and thinking in the abstract. It is the programs that we will develop and configure in the future that will control processes and generate the final forms.

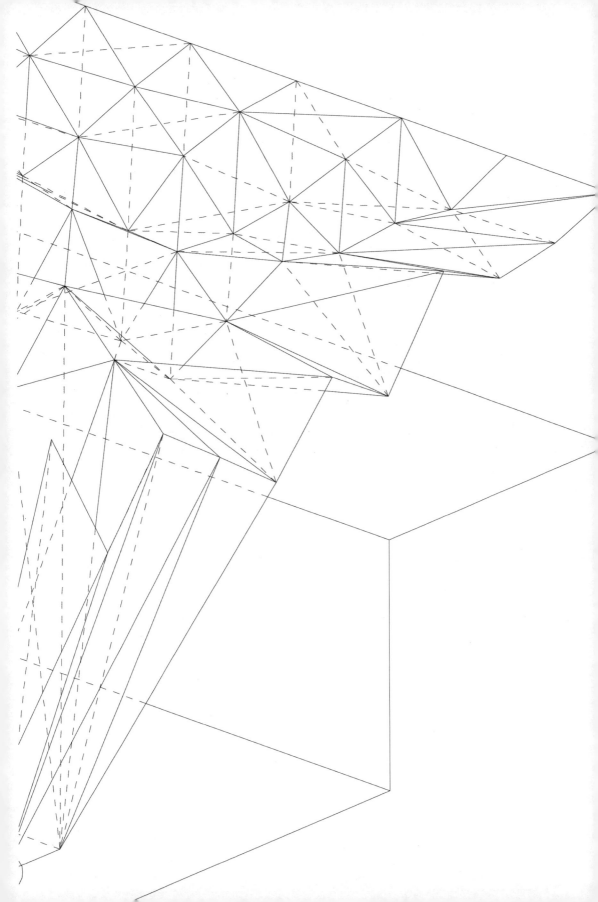

Ludwig Wittgenstein:
Philosophical Investigations

' This was our paradox:
No course of action could be determined by a rule,
because any course of action
can be made out to accord with the rule. '

建筑和体块

Buildings and Volumes

项　目：Grünhof(瑞士，苏黎世)

时　间：2004—2005

参与者：Alexander Lehnerer, Markus Braach

合作者：Kees Christiaanse Architects and Planners (KCAP) (Zürich, CH)

投影禁令 Forbidden Shadows

在苏黎世，有一条使高层建筑变得相当复杂的法规，即所谓的"两小时阴影规则"。这条法规规定：在一年当中的任意一天里，一座建筑物阻挡阳光到达其相邻建筑的时间不能超过2小时。显然，这样的条件对设计一栋高层建筑增加了额外的复杂性。每一个设计都必须回顾检查，看看其投射的阴影是否违反了上述法规。这个附加的证明工作通常都需要进行不止一次，这意味着设计过程会在一定程度上被延长。在传统设计方法下，设计工作者要面对这条法规所带来的问题而增加不合理的工作负担，因此产生一种状况，即法规规定者完全没有预见到高层建筑的设计变得很困难，在很多地区甚至是不可能的。

The city of Zurich has a building regulation that renders the building of high-rise blocks rather complex, the so-called 'two hour shadow rule': On any given day of the year, a building cannot obstruct the sunlight reaching its neighboring buildings for more than two hours. Quite obviously, such conditions add additional complexity to the design of a high-rise building. Every design must be checked in retrospect to see whether it casts shadows that contravene the above rule. This additional proofing work, often needing to be carried out more than once, means that the design process can be considerably protracted. Since, with traditional methods of design, one is often faced with an unjustifiable workload to get around the problems that this regulation poses, the situation arises—one not entirely unforeseen by the law-makers—where the building of high-rise blocks becomes extremely difficult and, in many places, practically impossible.

在苏黎世 Grünhof 项目中，我们做到了在不利的条件下也能提供不同类型的选择。与其他的项目相比，这个任务显得更为简单，因为我们只考虑阴影的影响。它是我们实验室工作方法的一个典型例子，因为结果清晰地证明了电脑辅助建筑设计带来的可行性。

With the Zurich Grünhof project, we were able to show the types of options that can be produced—even in unfavorable conditions. In comparison to other projects, the task may have been easier since we only had to calculate a shadow study. It was, however, a prototypical example for the working methods

Grünhof (Zürich, CH)

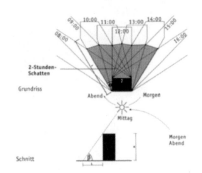

在苏黎世，建筑的体量除了受建筑允许高度和建筑用地所允许的发展限度限制，还受"阴影规则"的限制。此规则是防止建筑物的阴影在一天之内投射到其相邻建筑上的时间超过2小时。在设计建筑的过程中就需要不断地进行修改，以使其满足这个规则。如果是人工设计，这种修改需要很大的工作量。在苏黎世 Grünhof 项目中，我们已经将这个过程完全自动化了。

In the canton of Zurich, building volumes are not only regulated by the maximum allowed building height and the permissible development of a building plot. Additionally, a 'shadow rule' is applied that prohibits buildings from casting shadows on their neighbors for more than two hours per day. This often results in the need to keep working upon the designs until they conform to this rule. If performed manually, this always generates an immense amount of work. We have completely automated this approach for the Zurich Grünhof project.

#35

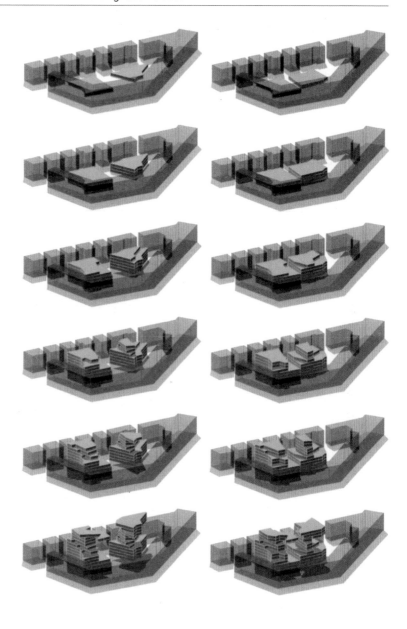

Grünhof (Zürich, CH)

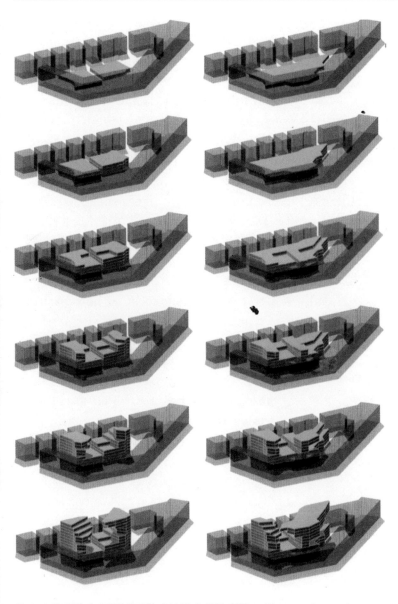

该项目的出发点是通过覆盖建筑综合体的庭院来增加建筑密度。从上往下所显示的是四个变体的生成过程。生成的规则是：一年当中的任何一天里，庭院中新建筑（绿色）的任何部分，投射在已有建筑（棕色）上的阴影都不超过2小时。

The starting point was to achieve an increased density by covering the courtyard of a building complex. The generation process of four variants is represented, reading from top to bottom. The rule becomes: No part of the new building (green) in the courtyard can cast shadow onto the already-existing building (brown) for longer than two hours on any given day of the year.

of our department, since the results clearly illustrated the possibilities enabled by computer-aided architecture.

该项目的出发点是将现有建筑综合体的内院进行延伸与连接。用传统的设计工具，先生成一个设计方案，再对这个方案进行阴影计算，根据数据进行方案修改，再进行阴影计算，以此类推。这是一种很不方便而且繁杂的工作方式。

The starting point was the extension and consolidation of the inner courtyard of an existing building complex. With classical planning tools, a design would have been produced, the shadows calculated, the design modified according to the first shadow study, the shadows recalculated, and so forth: an inconvenient and laborious method of working.

从草图中我们可以很快地看出，如果使用传统的规划方法，这个建筑用地上将不会出现特别高的建筑。然而，如果使用像 Kaisersrot 这样正确编程的软件，结果会是惊人的。用这个程序来计算阴影时，它并非是按照从头到尾的标准程序来进行的，而是反过来从阴影开始计算。从编程的角度来说，实现这种反转并不困难。实际上，这更像搭建乐高积木。电脑把一块积木放到另一块积木之上，一旦一块积木投射的阴影是不符合要求的，则这块积木就将被丢弃。保存"好的"积木，丢弃"坏的"积木，这是一个渐进的过程，也是我们的很多项目彰显的特点。与其他工程相比，这个特殊案例里计算的复杂程度大大地降低了，但其效率及说服力并未减弱。

We could see immediately from the sketches that on this plot there could be no particularly high buildings if traditional planning methods were employed. However, with an appropriately programmed piece of software like Kaisersrot, the results can be surprising. The application of the program when it is set up to calculate shadows doesn't work in the standard direction—from beginning to end—but rather the other way round, starting with the shadow. From a programming perspective, this reversal is not a difficult task; in fact, it's a bit like building with Lego bricks. The computer puts one brick on top of another, and as soon as one of the bricks casts unfavourable shadows it is discarded. The 'good' bricks stay, the 'bad' bricks disappear—an evolutionary process, something which marks many of our projects. In this particular case, the level of calculation complexity is greatly reduced when compared to other projects, but it is no less efficient and illustrative.

一旦我们从反方向着手计算，在建筑物升高时，程序总是能保证建筑物投射阴影的时间不会超出允许值。电脑

Since we commence calculating in reverse, as a building grows higher, the program constantly assures that it will not cast shadows

生成了建筑的最大允许体积——体积有的时候会很庞大，并且总是有着怪异的形式——它永远不可能用人工方法完成。然而，必须清楚电脑辅助设计得到的并不是最终的解决方案，而是作为建筑师工作的基础。在这个项目中，我们能够明显地看到，建筑师得到的不仅仅是能够代替他们做重复性工作的资源，还得到了进行深化设计的丰富基础，因为电脑为设计师提供了一个其他方法很难得到的设计框架。

如果认为 Kaisersrot 提供的是最好的或者是唯一可行的解决方案的话，那你就错了。在任何情况下，单一结果原则上对经验模型参数的评价都是无效的。Kaisersrot 给出了复杂边界条件框架内的可靠结果——其结果均是其他方法无法获得的。这个原则不但对苏黎世 Grünhof 这样的项目适用，对我们软件应用的整个领域都适用。一旦程序开始运行，计算过程就提供了潜在的新的解决方案。不能排除有更完美的结果，不过你可以随时打开程序继续计算以得到更好的方案。对建筑师来说，计算建筑阴影的过程一直是有点徒劳的苦役，而现在 Kaisersrot 软件把建筑师们从这个苦役中解放出来。

for longer than is permissible. The computer develops the maximum allowed volume—this volume is sometimes enormous and often of a bizarre form—and can never be arrived at using manual methods. However, it is important to understand that the computer-aided designs are not complete solutions, but rather serve as a basis for the work of the architect. In this project it was clear to us that the architects gained not only resources because they were able to delegate repetitive tasks; but they also obtained a rich basis for their further designs, since the computer delivered a design framework that would otherwise have remained inaccessible.

It would be a mistake to think that Kaisersrot delivers the best or the only possible solution. In any event, stand-alone results are, in principle, not valid for the evaluation of the parameters of empirical models. However, Kaisersrot delivers solid results within the framework of complex boundary conditions—results that would otherwise be unachievable. This principle is valid not only for a project like the Zurich Grünhof, but also for the entire area of application of our software. The calculation process delivers potentially new solutions as soon as it is run. It is impossible to rule out the fact that there might be an even better solution, but one can always rerun the software to look for better outcomes. For the architect, this process has always been a somewhat Sisyphean task, from which he now is relieved by the Kaisersrot software.

项　目：**Stadtraum Hauptbahnhof（瑞士，苏黎世）**

时　间：始于 2004

参与者：Alexander Lehnerer, Markus Braach

合作者：Kees Christiaanse Architects and Planners (KCAP) (Rotterdam, NL)

建筑作为调节器 Architecture as a Regulator

对中央车站周边的城市空间的重建工程，是苏黎世州最振奋人心的建筑项目的一部分。新的街区最终出现在轨道边，从设计的角度来说，这个区域的设计一直困难重重。在瑞士的建筑与设计杂志 Hochparterre 的特刊里，并无不公地将该地当时的发展历史描述为 "30 年的失误"。能提供 320 000 平方米的可用空间，为 1 200 位居住者提供住房，以及提供 8 000 个工作岗位，仅这些数字就足以说明，这是一个决定苏黎世城市进一步发展的重点项目。项目预计在 2018 年完成，其投资成本估计为 15 亿瑞士法郎。

The regeneration of the urban space around the Central Station forms a part of the most exciting building project in Zurich. A new city quarter is springing up on the edge of the tracks, after—from a design point of view—a site history beset with difficulties. In a special edition of Hochparterre, a Swiss architecture and design magazine, the then current development history of the site was described—not unjustly—as '30 years of blunders.' The figures alone were enough to tell you that this was a key project for the further development of the city of Zurich: up to 320,000m^2 of usable space was to be provided, housing for up to 1200 occupants and up to 8000 jobs. The investment costs for the project—slated to complete in 2018—were estimated to be 1.5 billion Swiss francs.

瑞士国家铁路公司举办了一场研讨会形式的竞赛，他们希望从这个竞赛中能够出现城市规划与发展的概念。所面临的关键性任务是权衡土地所有者、投资者以及城市代表们之间的利益。这包括建筑可能的高度与体积，以及新建城市空间与相邻区域间的联系等诸多方面。需考虑的众多方面简言之包括：建筑的连续性、灵活性及连通性，对公共空间的重新评估，建筑类型，工程分划，成本和潜在的投资回报率，以及生态和经济。

The Swiss national railway company set up a competition in the form of a workshop, from which they hoped an urban planning and development concept would arise. A crucial task up front was the mediation between the interests of the owners of the land and investors on the one hand, and those of the city representatives on the other. This included aspects such as potential building heights and volumes and the connection of the new urban space with neighboring areas. Key words of the brief included: continuity, flexibility and connectedness, re-evaluation of the public space, building typology and project division, costs and potential return on investment, as well as ecology and economy.

与荷兰建筑实践机构 KCAP 合作的这项工程，对我们来说是进一步将我们的工作方法及概念路径的潜能加以测试的机会。结果很成功，我们对于城市空间提出的一个开放的、动态的设计和建造过程的概念正在实现。

This project—on which we collaborated with the Dutch architecture practice KCAP—was for us a further opportunity to put the potential of our working methods and our conceptual approach to the test. It turned out to be a success, since the concept that we proposed for the urban space—one of an open, dynamic design and building process—is being implemented.

我们的出发点是确保新社区与周围城市建筑物之间的连续性，避免与环境形成强烈的反差。扩张现有的城市网格产生了各种各样的建筑用地。在讨论建筑的高度时出现了第一个分歧：投资者想要建筑越高越好，以使其每平方米建筑用地的回报率达到最大化。然而，特别是在苏黎世，高楼的建造总是颇受争议。这个城市最不想要的就是在火车站的地面上建一座小型的曼哈顿。

The starting point for our ideas was to ensure continuity between the development and the surrounding urban fabric and not to create a sharp contrast to the surroundings. Extending the existing city grid yielded the various building plots and sites. It was in discussions on the building heights that the first conflicts arose: The investors wanted to build as high as possible to maximize their per-square-meter return on the building land; whereas, particularly in Zurich, the building of high-rises has always been a subject of much debate. The last thing that the city wanted was a small Manhattan springing up on the grounds of the railway station.

解决这个常见的城市规划分歧，我们从两个阶段着手实行：首先，是手工建造的网状框架，确定建筑允许的最大高度和体积——外部界限的确定，不过是三维建筑红线的模型——这只能在特定的条件下进行讨论。其次是在这个限定下产生确定建筑物生成的内部编程和参数化规则系统。这包括

The solution at which we arrived for this common urban-planning conflict played out in two stages: Firstly, there are hand-constructed mesh frameworks that define the maximum allowed building volume and building height—the external definition of thresholds, little more than a matrix of three-dimensional building lines—that could only be broached under

在"Stadtraum HB"(火车站城市空间)一词是指在苏黎世中心一座废弃铁路车场生成新城市街区的城市设计项目。和 KCAP 一起,我们利用动态、开放的建造和发展过程来获得最大建筑空间的设计理念赢得了竞标。

Behind the term 'Stadtraum HB' (city space Main Station) hides an urban design project that is generating a new city quarter on a disused railway marshalling yard in the heart of Zurich. Together with KCAP, we won the competition with our concept of a dynamic and open building and development process that complies with the regulations on the maximum allowed building volume.

日照及阴影状况、居住空间与工作空间之间的比率、不被干扰的视线、向特定街道的开口、建筑的可寻度,以及建筑物之间近似的比例关系。最后这一点很有趣,因为它显示了设计的思考同样能够影响这样一个调整系统。在内部与外部调整系统的调和下,建筑最终生成,具有动态性,以过程为导向,且最后一秒都可以调整。通过使用描述总体规划的设计工具,这个过程将始终保持灵活可变。从外面看这像是一个束缚了建筑设计的笼子,实际上却为区域设计提供了更为宽阔的界面。

我们的设计的一个长处就是它事实上是一个实际的规划概念。在设计的每一步中,设计理念能够很轻松地被更新和确定。不同的可能性与地方规范相互作用,使得一个建筑单体的实践不需要提供整个设计。诸多实践可以同时致力于庞大而耗时的项目,同时不忽视项目的整体城市环境。因此一个设计和建造过程可以靠它自身发展,形成成熟的结构和 "自然" 的城市组织,不会在周边环境中插入怪异或突兀的结构;新的发展适应于已有的多层面的城市状态。程序

very particular circumstances. Secondly, there is an internal, programmed and parameterisible system of rules that define the possible growth of the buildings within this virtual envelope. This included sunlight and shadow conditions, the ratio between living space and working space, lines of sight that could not be interfered with, the openings to particular streets, the addressability of the buildings, and finally, the approximate proportional relationship of the buildings to each other. This last point is interesting in as much as it shows that design considerations can also influence such a regulatory system. In the conflict between the internal and external regulatory systems, architecture can now arise, dynamic, process-oriented, amenable to revision until the very last moment, and which—via the design tool represented by the master plan—remains as flexible as it is consistent. What looks on the outside like a cage, like something that would actually constrain the architecture, is, in reality, something that provides a comparatively broad canvas for the design of the area.

One strength of our design is that it is actually a tangible planning concept. At every step of the way the concept can be easily updated and finalized. Thanks to the interplay between copious possibilities and local specifications, it becomes unnecessary for a single architecture practice to deliver the entire design. Instead, several practices can work on a large and time-consuming project without losing sight of the overall urban context of the project. A design and building process can therefore

根据其内部和外部规则，运行时建造参数遵循相关背景及准则，即便有很多人在做这项工程，或者有新的投资者带着新的需求加入这项工程当中也无妨。花最小的精力，新的变体可以在任何一点被评估并且解决问题，例如：一个更高的塔，或者一个额外的塔，将会对整个设计的日照和阴影条件产生什么样的影响？（这对于独立的住宅区也同样适用）对于建筑的成本及可能回报又有什么影响？所有这些问题的答案都能以客观、易于理解的方式传递到项目的每一个参与者手中。

像 Kaisersrot 这样的软件允许建筑的发展保持动态与开放的状态，到最后一刻才会形成最终的结论。成本和实施过程可以在任何一个给定点进行调控。过程的动态性不仅改变了建筑的技术条件，同时也令所有参与者产生新的理解——包括投资者、城市官员以及建设管理人员。在现实世界中，人们不可能不用物质模型来展现最终目标，尽管这个过程只是对原则的一个说明，这种单一选择有别于从众多中择取。正如如何用木材来模拟一个动态过程？这只能用快照式的状态呈现。对于大家都想得到答案的一个问题，这是可以理解的形式，比如这项工程到底会变成什么样。然而在未来要给出一个有意义的答案不是那么容

be initiated that—practically all by itself—leads to a mature structure and a 'natural' city fabric. No alien or monolithic structure will be imposed upon the surroundings; the new development will fit into an existing, multifaceted city. The program, with its external and internal rules, runs in the background and guarantees that the building parameters are adhered to—even when many people are working on the project, or when new investors with new demands come on board. With minimal effort, fresh variants can, at any point, be evaluated, and problems solved: What effect would a higher tower, or an additional tower, have upon the sunlight and shadow conditions of the entire development — and also on the individual dwellings — and what effect would it have on the building costs and the potential income? All these answers can be delivered to the participants in an objective, comprehensible manner.

Software like Kaisersrot allows the building development to be kept dynamic and open, the final solution crystallizing at the last moment. The costs and the implementation of the process can be controlled at any given point. The dynamic nature of the process changes not only the technological conditions of building, but also demands and generates a new understanding among all those taking part — including investors, city officials, and building managers. In a real world scenario, obviously, one cannot do without a physical model for presentation purposes, although in this process, it can only really be an illustration of principle, an arbitrarily chosen variant from among many.

Stadtraum Hauptbahnhof (Zürich, CH)

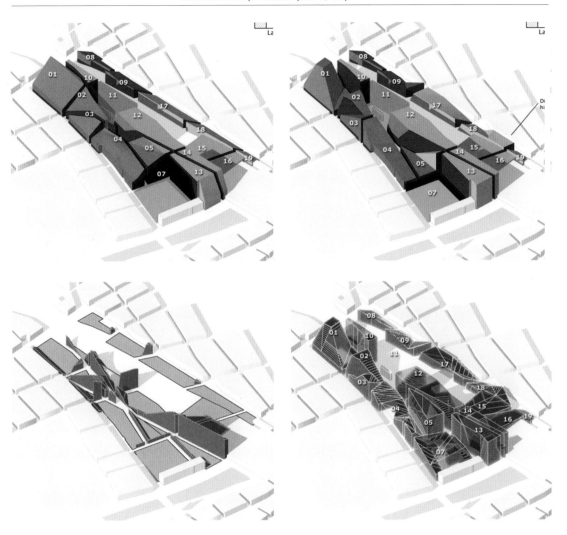

最大的建筑外轮廓体积（上左）在一系列规则下生成，这些规则组织光线与阴影的比率、视线、场地的开发以及建筑比例之间的关系。整个系统显得不确定，或者说是"具体前状态"，但不损害整个系统的功能。

The maximal building shell volume (top left) worked in conjunction with a set of rules that organized the ratio of light and shadow, lines of sight, site development, and relationships between the built proportions. The appearance of the total system becomes undefined, or rather 'prespecific,' without damaging the function of the total system.

对于建筑外轮廓生成的多个方案。　　　　　　Variant studies for the generation of the building shell.

易的，这一点我们是确信的，并且，这些答案不会像直到现在人们还习惯的那样，在建造过程中很早出现。我们仍然不知道这个新的城区会是什么样子，但我们知道，它将在规定的框架条件下运作。

Just how does one model a dynamic progression in wood? It can only ever represent a snapshot. To pose the question that everyone wants answered—how the project will actually look—is, of course, understandable. However, in the future—and of this we are convinced—it will not be so easy to give a meaningful answer, and neither will the answer come as early in the building process as people have been used to until now. We still do not know what this new city quarter will look like, but we know that it will function within the stated framework conditions.

一个城市永远都不会真正地完成：它的发展始终是一个不间断的过程。通过我们的电脑辅助及驱动的方法，我们试图尽可能真实与可靠地表现这种开放性。当一个建筑学意义上很特殊的新的城市街区在苏黎世中央火车站附近拔地而起，我们的目标就达到了。但是，究竟要有两座、三座或者是四座大楼矗立在那里，到目前为止还无法回答。

A city is never truly completed: its development is always an ongoing process. With our computer-aided and -driven methods, we are trying to represent this openness as solidly and as authentically as possible. We will have finally reached our goal when an architectonically differentiated, new city quarter arises near Zurich's central railway station, connected to the existing one. But the question of whether two, three, or four towers will be standing there is, for the moment, unanswerable.

项　目：	**Bishopsgate**（英国，伦敦）
时　间：	2004
参与者：	Alexander Lehnerer
合作者：	Kees Christiaanse Architects and Planners (KCAP) (Rotterdam, NL)

城市的谈判 Cities Negotiating

回顾伦敦的Bishopsgate项目，其本质与苏黎世中央车站的项目很相似，虽然规模上前者比后者要大很多。像在苏黎世一样，伦敦的开发场地同样占据了城市的一个中心地段，原为城市中心区边缘的铁路货场是这个大都市首要开发区最后的自由地之一。这个地块几年前就已提出了很多的规划方案，它们要么都被搁置一边，要么因与股权持有者的利益产生冲突而被撕毁。像伦敦这样的城市，对开发区每平方米的投资回报看得比苏黎世还重要，经验证明这不是对建筑质量有利的影响因素。

与苏黎世铁路工程风格相似，我们不希望我们的研究产生最终的设计结果，而是运行一个程序，使之表达并阐明各种相互矛盾的因素对场地的影响。我们又一次与KCAP的设计工作室合作。

在这个例子中，同样也需要采用特定的规则进行定义，并且将它们整合到设计中去。例如，在圣保罗教堂旁边，保持一条清晰的视线非常重要，确定最初的建筑用地，保证整个场地的通透性和循环组织。每

Looking at the brief, the project in Bishopsgate in London was essentially similar to the project at the Zurich railway station, albeit on a considerably larger scale. As in Zurich, the London development site also occupies a central location in the city, a former goods railway yard on the outskirts of the center, and one of the last free parcels of prime development land in the metropolis. Many planning proposals had been submitted for it in previous years, and had been shelved, torn apart by conflicting interests of the stake-holders. In a city like London, the return on investment per square meter is much more of a central concern for a development than in Zurich—a factor that experience has shown not to be very conducive to architectural quality.

In a similar vein to the Zurich railway project, we did not want our study—again carried out in cooperation with the planning office of KCAP—to produce a final design, but rather to set in motion a process that would articulate and illustrate the various competing factors affecting the site.

In this case, too, it was necessary to adopt certain rules, define them, and integrate them into the planning. For instance, it was crucial to maintain a clear line of sight to nearby St.Paul's Cathedral, and to define the original building plots to guarantee the permeability

座塔楼的选址都需要确认，也需要保证周围建筑在特定的视野内不会出现庞大的体量。同样也要考虑对日照和阴影的研究，虽然在伦敦没有像苏黎世那样的"2小时规定"。

这些相对简单的规则（总共有8条）都被编入程序形成了设计的基础。就像在苏黎世，我们主要关心的是如何开发充满冲突的建筑面积，而不采用任何生硬快速的经验性建议。在我们的经验当中，这些做法很有可能从一方面或另一方面影响到居民。虽然不能立竿见影，但是我们规定软件遵循的规则使建筑在既定的框架条件内有了更多的喘息空间。这些游戏的规则使得所有参与者都能扮演有意义的角色。

从某种意义上看，城市规划的作用退居二线了，在不影响设计的情况下，它定义场地规则。我们对形式并不感兴趣，而是对不同功能之间的交互感兴趣。这也许是当代建筑与城市规划所不熟悉的。然而，我们研究的不是美学的问题，而是一个可自由发挥的过程。如果这些过程的结果是对冲突利益之间更好的整合的话，我们就已经达到目的了，因为我们产生的众多有差别的结果同样会产生更好的设计。

and the circulation logistics over the entire site. The siting of each tower would need to be confirmed, and certain views onto the surrounding buildings conserved in order that the complex not appear monolithic. The sun and shadow studies were also a consideration, although in London there is no two-hour rule as in Zurich.

These relatively simple rules—eight of them altogether—were programmed in and formed the basis of the planning. As in Zurich, our primary concern was the opening up of the conflict-laden square, without making any a priori hard-and-fast recommendations—something which, in our experience, is likely to incite resistance from one side or the other. Although not immediately evident, the rules defined by us and followed by the software give the architecture more breathing space within the stated framework conditions. They are the rules of a game that ensure that all involved can play a meaningful part.

In this sense, urban planning takes a back seat, conforming to defined ground rules without trying to influence the design. We are not interested in the form; rather, we are interested in the interplay of the various functions. This may be something unfamiliar in contemporary architecture and urban planning. However, the latter here is not primarily a question of formal aesthetics, but rather one of processes that one may unleash. If these processes result in nothing but better integration of the conflicting interests, we will already have achieved a great deal, since the many differentiated results that we can produce also allow for a better design.

#39

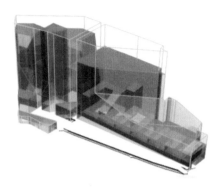

允许的建筑高度以及建筑之间的空间将会生成最大的允许建筑体积。

The permitted building heights and the space between the buildings will generate the maximum allowed building volume.

#40

 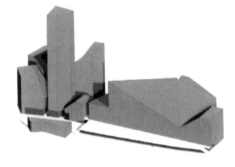

重要的视线将不会受到建筑阻挡。
为了更方便地布局，城市景观中的建筑单体如图所示。

Important lines of sight will not be interrupted by the building volume.
The individual buildings will be perceivable as such in the city landscape, in order to make locating them easier.

#41

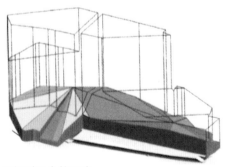 

最底层得到最少的阳光。
对临近建筑的阴影效应限制了建筑物的高度。

The lowest floors deserve a minimum of daylight. The shadowing effect on neighboring buildings limits the building height.

Bishopsgate (London, UK)

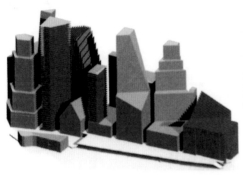
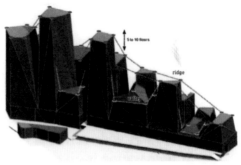

#42

建筑表皮将会被多样化,以避免城市景观的单调。
与山丘地形相似,不同高度的塔楼会相互分开,并根据其高度进行组合。

The volumes of the building shells will be diversified in order to avoid a monotonous cityscape.
Similar to a hilly topography, the differing-height towers will be offset in relation to one another and grouped according to their heights.

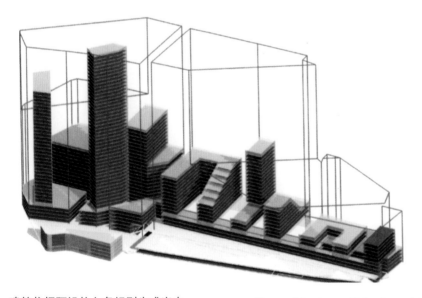

#43

建筑依据预设的七条规则生成出来。

The architecture will be formulated according to the seven preceding rules.

Gertrude Stein:
Tender Buttons

' Act so that there is no use in a center.
A wide action is not a width.
A preparation is given to the ones preparing.
They do not eat who mention silver and sweet.
There was an occupation. '

通用元素

Generic Elements

项　目：**Replay Column Atlas**

时　间：2003—2004

参与者：Kathrina Bosth, Markus Braach, Susanne Schumacher

学　生：Roman Brantschen, Sebastian Engelhorn, Patrik Hass, Meike Kniphals, Simon Mahringer, Thomas Maiss, Jessica Mentz, Niklas Naehrig, Niklas Reinink, Fiona Scherkamp, Martin Sigmund

合作者：Kunsthistorisches Institut der Universität Zürich (CH) – Prof. Dr. Hubertus Günther

现代古迹 Modern Antiquity

结构、内容和形式的分离是现在信息技术的基础支柱之一：随着因特网标准 HTML 的普及，XML 形式（可扩展标记语言）成为突出的例子。结构、内容和形式的分离也是我们工作室承担的很多项目的基本原则。尽管我们的工作重点在于为设计和建造中的数字技术，但是"柱式重构"计划——一个基于 XML 的关于最重要的经典柱式的概览——也是我们研究中非常重要的一部分。

The separation of structure, content, and form is one of the founding pillars of modern information technology: as a generalization of the Internet standard HTML, the XML format (Extensible Mark-up Language) is a prominent example. The separation of structure, content, and form is also a founding principle of many projects undertaken in our department. Although the main emphasis of our work lies in establishing digital technologies in the practices of design and construction, the 'Replay Column Atlas' project—an XML-based overview of the most important treatises on classical capitals—was a very important piece of work for our field of study.

柱式和柱头是建筑史中很明确的内容；它们在历史上的排序和分类是建筑理论的发展中最重要的基石，联结着今天和 2000 年的建筑历史。以论文的形式，建筑师和理论家们已经分析了很多历史上的例子，对它们的元素进行了分类和系统化。经过几个世纪，不同的表现方式被发展出来，图解技术、数学、几何学以及一个又一个的理论观点交替出现。在过去几年里，表达的复杂性呈现增长的趋势。而且几何结构的交流日益由文本向图形转换。

Columns and capitals may, studied in isolation, be a straight forward phenomenon from architectural history; however, their historical ordering and classification counts as one of the most important consolidations in the development of architectural theory, connecting the present day with 2000 years of building history. In the form of treatises, architects and theoreticians have analyzed examples from history, cataloguing and systematizing their elements. Over several centuries, different representational conventions were developed, in which

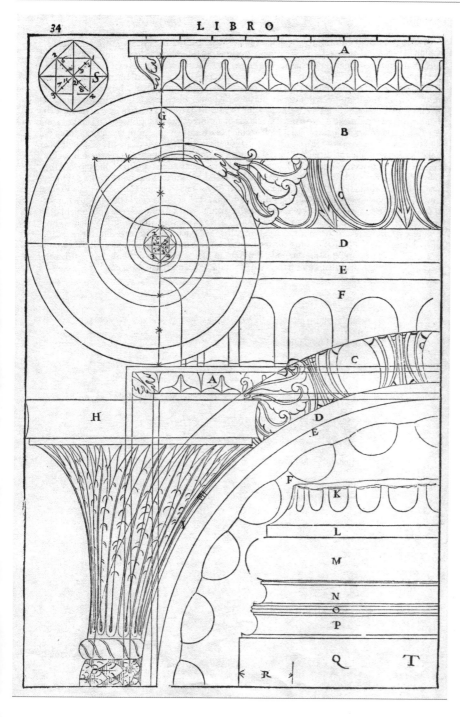

Illustration from: Andrea Palladio. I quattro libri dell' architettura (1570).

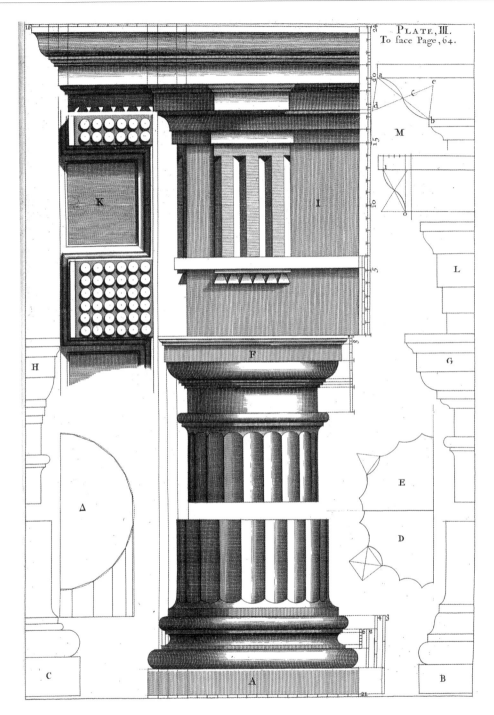

Illustration from: Claude Perrault, Ordonance descinq espèces de colonnes selon la méthode des anciens (1638).

illustration techniques, mathematics, geometry, and theoretical standpoints play off one another. Over the years, the complexity of the representation has tended to increase, and the communication of the geometric construction has shifted increasingly from text to pictures.

"柱式重构"这一项目想法的产生，是因为各种柱式高度模式化，可以把柱式转化成为类似 XML 这样标准化的现代格式。这项工作的基础是关于多立克和爱奥尼柱式的《建筑十书》。选择这些论著是因为它们贯穿了尽可能多的时期，并包含了在设计和形式方面最广泛的种类，从古典的维特鲁威的工作、贯穿文艺复兴时期的著作，到 Claude Perrault 和 Jean Nicolas Louis Durand 二人的研究，他们都建立了更抽象、更普遍的体系，再到那些以遗址的实际测量为基础的著作，最终到 Robert Chitham 的 20 世纪以来的建筑学古典柱式。用结构、内容、形式三者分离的原理对所有例子进行严格的分析之后，我们用三个步骤把古典案例转化为现代 IT 格式。第一阶段——主要的形制是写成 XML 文件，即众所周知的模板。第二阶段是利用文件来组织精确定义的个体元素的形体。最后一个阶段，最终的输出形式能够根据最终使用者的需要来选择。在这种情况下，我们选择图像表现形式，经由 SVG（可缩放矢量图形）格式，它能够比较历史图解。因此表格、文本以及三维的输出形式都可以被考虑。

The idea for the project 'Replay Column Atlas' came about because the strong for-malization of the various orders makes it possible to transfer them into a standardized modern format like XML. The basis for the work was ten treatises on the Doric and Ionic orders. These treatises were chosen because they spanned the longest possible time period and included the greatest variations in design and form, from the classical work of Vitruvius, through to Renaissance works, to studies by Claude Perrault and Jean Nicolas Louis Durand, both of whom developed more abstract, general systems, on to works which were based on actual measurements taken from the ruins themselves, and finally to Robert Chitham's The Classical Orders of Architecture, from the 20th century. The conversion of these classical examples into modern IT formats was achieved in three phases, after analyzing all the works in strict adherence to the principle of the separation of structure, content and form. The first step—which held for all ten works—saw the general schema being written into an XML document, known as the template. The next step saw the use of sheets that held the precise and well-defined geometry of the individual capital elements. As a last step, the final output format could be chosen according to the end user's needs. In this case, we chose graphical

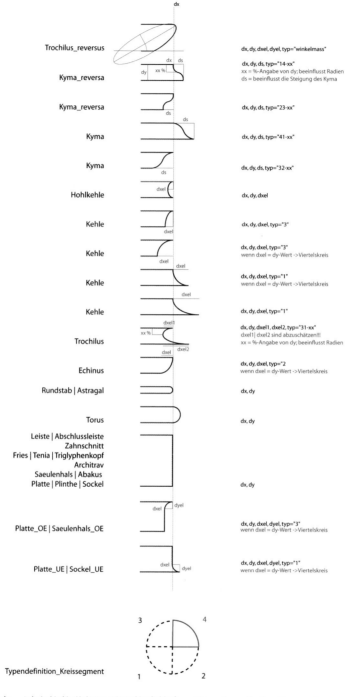

对于我们而言，过去的数世纪里重要的建筑中的柱式在语言上和几何上的精确描述，是 XML（可扩展标记语言）编码的起始点。

The linguistically and geometrically precise formulation of the column classes in the important architecture tracts from the various centuries was, for us, the starting point for a comprehensive encoding in XML (Extensible Markup Language).

Replay Column Atlas

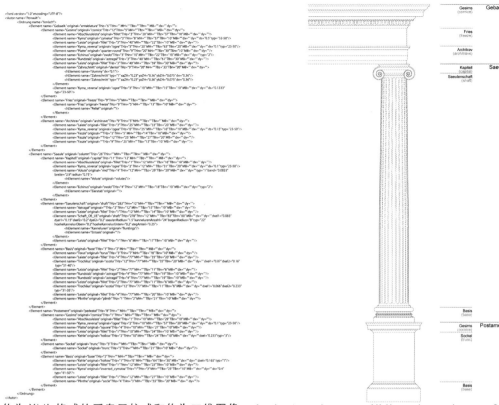

作为 XML 格式的爱奥尼柱式和作为二维图像的爱奥尼柱式。

An Ionic column as XML code and as a 2-D rendering.

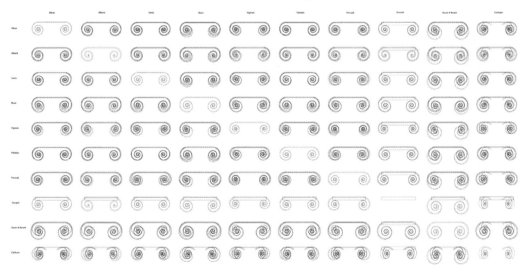

不同的建筑论著描述的爱奥尼涡卷非常不同。用 XML 对该柱式进行建模允许图像化的比较。

The different architectural treatises describe an Ionic volute, for example, very differently. The unified modelling in XML permits a graphical comparison.

representation via the SVG (Scalable Vector Graphics) format, which would permit a comparison with the historical illustrations. Tabular, text-based, and three-dimensional output formats could also be considered.

把风格截然不同的、时代不同的理论进行转化并统一起来，为比较和分析建筑理论提供了新的可能性。直到现在，以下论著还没有在视觉上的直观描述：维特鲁威的著作没有包含图，而文艺复兴时期的建筑师阿尔伯蒂也故意不加图释。在"柱式图样"中，一种新的论著已经变得可行，它能够提供并整合、比较建筑十书中的柱式，从最大的尺度到最小的细节。"柱式重构"能够延伸出新的论著，并且可以以三维形式建立数据库。

The conversion and consistent formatting of theories that had been put together in completely different styles, and in completely different time periods, opened up new possibilities for the comparison and analysis of architectural history. Until now, a visual representation of the following treatises had not been possible: Vitruvius's work contained no illustrations and the Renaissance architect Leon Battista Alberti had also purposely forgone them. With the Column Atlas, a new treatise was now available, one that provided an integrated comparison of the ten treatises under consideration, from the largest scale down to the smallest detail. The 'Replay Column Atlas' can be extended with new treatises, and the data is available in three-dimensional form.

项目的一个新奇之处源于简洁的特殊视角。从进行这项转移工作的人的视角来看，这些论著被有区别地分析，并且引发了不同的问题：信息是怎样被排列的？秩序是怎样建立的？这些描述和方法的代表——绘图或文字——从属于哪个历史观点？对这一主题的一个古典的纵向考察提供了对各个著作前所未有的深入理解。另一方面，"柱式图样"的一个特殊特色在于它对于论著的横向理解。它开启了用一种宏语言在同一层面上描述所有论著的途径，它可以独立地或整体地对所有论著进行对比和理解。

One of the novel aspects of this project was the special perspective that arose from the brief. From the point of view of the people responsible for the conversion work, these treatises are analyzed differently, and they raise different questions: how is the information arranged, how were the orders developed, which descriptions and methods of representation—drawn or written—pertain to which points in history? A classically vertical examination of a theme provides an ever deeper understanding of the individual work. A particular feature of the Column Atlas, on the other hand, is its horizontal reading of the treatise. It starts by describing all treatises on

建筑柱式十书的编码（高级）比较　　The code (high-level) of the ten column treatises compared.

通用元素 Generic Elements

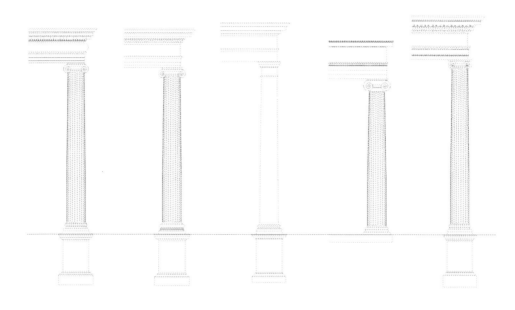

各种不同柱式的比较与相关程序编码的对比图表。

A graphic comparison of various column orders with a comparison of the relevant program code.

the same level using a meta-language, an approach which permits a differentiated understanding of the works, both individually and together.

从技术上讲，像苏黎世 Hardturm 这样实际的建筑项目与"柱式图样"这种研究建筑史核心课题的项目之间并没有本质上的差异，这一点很具有启发性。

Furthermore, it is an appealing thought that between a real-world building project like the Zurich Hardturm and a theoretical project like the Column Atlas, which deals with a central theme in architectural history, there is—from a techno logical point of view—no considerable difference.

#52

结构与形式的分离

有人可能这样认为，形容一把椅子是由什么构成的很容易——有四条腿，一个座，一个靠背，腿上还可能有脚轮或者座上还有坐垫。这些单独的元素是否能如我们预期的那样组成一个实际的座具，依赖于它们以一种正确的秩序被组合在一起，或依赖于它们组成部分的变形（见第二章"超越网格"）。然而，一把由 Stefan Wewerka 设计的椅子包含了组成椅子所需要的一切，但是没有人能够坐在上面，因为它的结构已经和外部形式分离了，它的拓扑关系和有形的几何体分离了。所以 Stefan Wewerka 椅子是一个例子，告诉我们怎样把沙利文的那个经常被引用的格言"形式跟随功能"应用在未来前景中。

结构、内容和形式的这种分离，在信息技术中是一个关于建模过程的通用方法。首先要有一个结构来容纳特定的内容。然后这个内容根据既有的环境被塑造成特定的形式。一个 MP3 播放器的形式基本上能够自由地和任意地被选择——不管是作为特别订制的电脑上的接口（例如 iTunes），或是在桌面的一个硬件（例如 iPod），两个都是同一内容的不同表现形式：在这种情况下是 MP3 播放器的标准数据库格式这个内容的不同表现形式。

这种原则同样能应用到建筑学中。我们一旦已经获得了一个与建筑物相符合的数据库结构——或者关于古典柱式的资料——无论这些数据在电脑显示器上以三维形式被壮观地呈现出来，或是作为一个组件明细被打印出来，还是以 1∶25 的比例、甚至是用木材以 1∶1 的比例由数控机床切割出来，都没有区别。因为不管采用什么具体形式，它们都按同一数据被表现出来，所以它们没有本质上的区别。既然这种成为二进制代码的转变允许任意数量的各种转换，那么从单一的拓扑关系建立最广泛种类的几何形式和从单一的结构建立最广泛种类的形式就成为可能，而所有的这些是由电脑控制的程序引起的。这种结构和形式的分离同样意味着不同的结构能够锁定一个形式，然而，这种逆转在建筑学中起到的作用还很小。

一旦结构和形式被明确地分离，并且它们之间形成清晰的接口，在工程师和建筑师之间经常充满的紧张状态的关系会变得缓和。直到现在，建筑师只是设计了外壳（无可否认的一种夸大），随后工程师试着为它设计一个结构。现在建筑师和工程师能够精确地几乎是彼此独立地建立外壳和结构，而不会导致任何重大问题而进行最终调整。在我们的模式中，建筑师主要对形式的表现起作用——采用 CAD，而工程师主要对拓扑关系的结构起作用——不采用 CAD，在工程领域这是一个令人惊奇的转变。像前面描述的那样，运算法则能够把工程师的拓扑关系转化成为由建筑师设想出来的形式。

建筑的角色也将相应地改变。现在，建筑学主要关注外部形式方面的内容。在过去几年里，由于计算机的应用，与"自由"形式有关的发展取得了巨大的进展。但是支持这些进展的必要结构和基础设施却相对滞后。开创性的建筑学因此是极其昂贵，有技术风险，并且经常是表现不佳的。我们正在促进形式和结构的分离——或者甚至说是结构、内容和形式三者的分离，使之成为相对独立的设计领域,并且能够证明在设计领域中大范围的自行整合是可行的。这些自动程序绝不会限制现在建筑师和工程师的创造力,相反,它们给予他们更多的自由去单纯地专注于他们自己的专业领域。这种结构和形式的分离使普通建筑物的建造也能得到仅在一些开创性的设计中才能见到的关怀和专注。或许我们正在见证一项特殊的瑞士建筑学的创新性进展,并在现实的建筑中起到关键的作用。

The Separation of Structure and Form

One would think that it would be easy to describe what a chair is made of: a chair has four legs, a seat, a back, and maybe castors on the legs or a cushion to sit on. Whether these individual elements could actually produce a seating opportunity—which is what one would normally expect—depends upon them being put together in the correct order, or, referring back to text II (Beyond the Grid → pp. 61f.) upon the deformation of their constituent parts. However, a chair like the one by the artist Stefan Wewerka, consists of everything that goes to make up a chair—but no one could ever sit on it, because its structure and its topology have been separated from the external form, from the tangible geometry. And so, Wewerka's chair is an example of how we can put Louis Sullivan's oft-cited aphorism 'form follows function' into perspective.

This separation of structure, content, and form is a commonly used method for modeling processes in information technology. First, there is a general structural kernel that is filled with specific content. This content is then moulded into a specific form, according to the precise context. The form of an MP3 player can, basically, be freely and arbitrarily chosen—whether as an individually tailored interface on a computer (iTunes, for example) or as hardware lying on a tabletop (an iPod, for example). Both are different ways of representing the same content: in this case, the standard data format (MP3) for an MP3 player.

This principle can also be applied to architecture. Once we have obtained a consistent data structure for a building—or treatises on classical column capitals—it is irrelevant whether these data are rendered spectacularly in 3-D on a computer screen, printed out as a parts schedule or turned into a 1:25-scale—or even a 1:1-scale—wooden model by a CNC milling machine. There is basically no difference, because they are all represented by the same data, no matter what concrete form it takes. And since the conversion into binary code permits any number of arbitrary transformations, it is possible to develop the widest variety of geometries from a single topology and the widest possible variety of forms from a single structure—all of these arising from computer-controlled programmes. The separation of structure and form also means that different structures can be locked to one form; however, this reversal only plays a minor role in architecture.

As soon as structure and form are clearly separated, and a clean interface between them created, the often tension-filled relationship between engineers and architects becomes considerably less tense. Until now, architects have often only designed the shell (admittedly, an exaggeration), with engineers subsequently trying to design a structure for it. Now, architects and engineers are able to accurately develop shells and structures largely independently of one another, without the final mutual adjustments creating any big problems. In our schema, architects work mainly with formal representations—with CAD—whereas engineers work mainly with topological structures—without CAD—an astonishing turnaround in the field of engineering. As described earlier, algorithms are able to turn the topologies of the engineers into the forms envisaged by the architects.

The role of architecture will change accordingly. Today, architecture is mostly concerned with external, formal aspects. In the last few years, thanks to the use of computers, enormous strides have been made regarding the development of 'freer' forms. But the structures and infrastructures necessary to support these have been increasingly lagging behind. Groundbreaking architecture is, therefore, extremely expensive, technologically risky, and often performs poorly. We are promoting the separation of structure and form—or even a three-way separation of structure, content, and form—as relatively autonomous fields of design, and are able to demonstrate that a largely automatic convergence of these fields of design is possible. These automatic procedures in no way limit the creativity of the architects and engineers. Instead, they give them more freedom to concentrate solely on their area of expertise. The separation of structure and form makes it possible for the structures of ordinary buildings to be produced with the care and attention that we have up until now only seen in a few groundbreaking designs. So maybe we are seeing a specific Swiss contribution to the development of architecture that, in real-world building, has a special significance for architectural design.

Roland Barthes:
Empire of Signs

' The streets of this city have no names.
There is of course a written address, but it has only a postal value,
it refers to a plan (by districts and by blocks, in no way geometric),
knowledge of which is accessible to the postman, not the visitor:
the largest city in the world is practically unclassified,
the spaces which compose it in detail are unnamed.
This domiciliary obliteration seems inconvenient to those (like us)
who have been used to asserting that the most practical is always the most rational [...].
Tokyo meanwhile reminds us that the rational is merely one system among others.
For there to be a mastery of the real [...], it suffices that there be a system,
even if this system is apparently illogical, uselessly complicated, curiously disparate:
a good bricolage can not only work for a very long time, as we know;
it can also satisfy millions of inhabitants inured, furthermore,
to all the perfections of technological civilization. '

建造

Construction

项 目：Olympia Stadiun（中国，北京）

时 间：2003

参与者：Markus Braach, Oliver Fritz

合作者：Herzog & de Meuron (Basel, CH)

尝试及错误 Trial and Error

"鸟巢"是北京奥运会壮观的国家体育场的著名别名。它是由瑞士建筑师赫尔佐格和德梅隆事务所设计的。其屋顶结构中看似随意排列的钢梁说明数字模型表现出的具体可能性与利用现有材料建造的可行性之间存在差距：任何在Photoshop图形处理软件中可以轻易组合到一起的构件，都会在绘图阶段和建造现场引起一系列的麻烦。

投标竞赛时鸟巢的正立面图是在Photoshop中绘制的。但中标后，必须建造出真实的立面，因此对当时的CAD绘图软件提出了无法解决的难题。由于结构设计和装配技术方面的要求，钢梁间的间距必须在某一确定的范围内：既不能比标准尺寸大，也不能比它小，并且钢梁交错产生的确切角度必须得以保留。就像前面所提到的，这些要求在Photoshop绘制的正立面中没有任何的问题，但是对于一个真实建造的且与众多钢梁定位密切相关的正立面来说就行不通了。不规则排列的钢梁意味着上述问题几乎无法仅凭手工解决。一旦在某一立面上的钢梁缺口闭合了，其他立面上的钢梁缺口就会张开，这就像是一个有着太多移动部件的谜团，每一个零件的变动都会影响到其他部件。

'The Bird's Nest' is the popular term for Beijing's spectacular Olympic Stadium, designed by the Swiss architects Herzog & de Meuron. The seemingly randomly ordered steel beams of the roof construction are an illustrative example of the gap between the representational possibilities of a digital model and the feasibility of construction using current materials: What can be put together with comparative ease in Photoshop can lead to a variety of problems on both the drawing board and the construction site.

For the competition, the front elevation of the Bird's Nest was composited in Photoshop. With the competition won, the rear elevation had to be constructed—something that posed an insoluble problem for the then-current CAD tools. For structural engineering reasons, and because of the assembly technology, the openings between the steel beams had to fit within a certain size range: neither larger nor smaller than a certain dimension, and certain angles where the beams crossed had to be preserved. As previously said, for the front elevation in Photoshop, this was no problem. However, for the rear elevation—which is tied intimately to the positions of the beams in the front elevation, this was not an option. The irregular ordering of the beams meant that this problem simply could not be solved by hand. As soon as a gap was closed on one elevation, a gap would open up on the other—it was like a puzzle with far too many moving parts, each one influencing the others.

Although it was relatively easy to program, the study, carried out in cooperation with Herzog & de Meuron, remains, to this day, one of the defining projects of our program of research. We were not trying to draw the solution; we were actually trying to develop an evolutionary mechanism that would deliver the solution to us. We started off with the 'mother of all solutions': a completely chaotic ordering of the beams. We took, for example, 100 beams, giving us a genetic code for the solution with 600 variables, since the position of each beam can be exactly described by coordinates in x, y, and z and the rotation of each beam by angles of x, y, and z. The first, completely random ordering of the beams threw up a catastrophically high number of errors. We then generated five copies of this solution. However, each copy differed slightly from the original, because we had randomly changed the values of the variables and thus the positions of the corresponding beams. In a manner of speaking, we had let the genetic code 'mutate,' and these mutated versions generated solutions with fewer problems. The solution with the fewest problems was copied/mutated anew, and we again chose the best solutions that were generated ... copy/mutate/select ... copy/mutate/ select... after 600 generations,

虽然对于程序而言这是比较容易处理的问题，但直到今天与赫尔佐格和德梅隆合作发起的这项研究仍然是一个标志性研究项目。我们不是试图绘制出解决方案，而是尝试创建一种能生成解决方案的进化原理。我们从所有问题的根源处即一堆完全混乱排列的钢梁着手。以100根钢梁为例，给出了包含600个参数的遗传编码方案，因为每根钢梁都可以通过x、y、z坐标和x、y、z坐标的旋转角度被精确定位。刚开始，完全随机组合的钢梁中产生了大量灾难性的错误。接下来，我们从这个结果衍生出5种变体。但是，每个变体都与初始状态稍有不同。因为我们随机改变了参数的值，这样就改变了交叉着的钢梁的位置。在某种程度上来说，我们令遗传编码突变，这些突变编码所生成的结果含有较少的错误。含有最少错误的方案被复制，重新突变、生成，我们再次选择所生成的最好的解决方案……复制、突变、选择……复制、突变、选择……直到600代后，我们找到了一个没有丝毫错误的解决方案。你可能会说我们将复杂的屋顶结构"晃动"了太长时间，然后让每个钢梁处于它们应该待的地方。这样我们最终得到了正确的解决方案。

北京奥林匹克体育场在解决问题的演化过程中不同阶段的视图。这项任务是要在钢梁放置的过程中使它们之间留有既不太大也不太小的缺口（用红色标出的地方）。从一个随机的排列开始（如左上图所示），通过遗传变异到600代后，最终进化算法生成了一个使各个钢梁位置都无任何问题的解决方案。

A view of the Beijing Olympic Stadium during various phases of the evolutionary problem-solving process. The task was to place the steel beams so gaps between them were neither too large nor too small (marked in red). Starting with a random ordering (above left) and using inheritance and mutations over 600 generations, an evolutionary algorithm eventually proposed positions for the beams to generate a problem-free solution (bottom right).

we had found a solution that exhibited no problems whatsoever. You could say that we had 'shaken' the complicated roof construction for so long, letting the individual members fall where they may, that we finally arrived at the correct solution.

事后，令人惊讶的是程序会产生一个仅有100条线组成的解决方案，但是我们无法用手绘出，也无法用传统的或者参数化的CAD软件绘制它。这就是我们一直所要寻找的方法和解决方案，正如我们说的不想把建筑置于计算机中，而是想通过计算机去梳理建筑。 正是因为能用很少的功夫就证实 game-changing 方法，鸟巢才在我们所有的项目中占据了如此突出的标志性地位。

In hindsight, it is astonishing that there can be a problem the solution of which consists of only 100 lines, but which cannot be drawn by hand, or with conventional or even parametric CAD systems. These are the briefs and the solutions we are looking for when we say that we don't want to put architecture on the computer, we want to tease it out of the computer. It is because we were able to demonstrate this game-changing approach with so little effort that, out of all of our projects, the Birds 'Nest occupies such a prominent and iconographic position.

| 116 建造 | Construction |

项　目：**Futuropolis（瑞士，圣加伦）**

时　间：2005

参与者：Christoph Schindler, Fabian Scheurer, Markus Braach

合作者：Studio Daniel Libeskind (New York, USA) – Thore Garbers;
　　　　Bach Heiden AG (Heiden, CH) – Franz Roman Bach, Hansueli Dumelin

助　理：HSG Universität St. Gallen (CH) – Holm Keller, Dr. Timon Beyes

赞助商：(Düsseldorf, D)

数字链效应 Digital Chain Reaction

从很多方面来看，这个项目都是我们对北京奥运会体育场研究的延续。但是它在技术上远比前者复杂，并且包含有一个决定性的进步：在 Futuropolis 项目中，完整的数字链形成，数字化设计和数字化建设生产间架起了联系的桥梁。这是我们研究中的重点建设项目，因为它不仅展示了在计算机辅助方法下建筑建造的可能性，而且还能通过一条跨越设计和生产的信息链获得经济效益。

In many ways, this project was the continuation of our study for the Beijing Olympic Stadium. However, it was technically far more complex, and included a decisive further step: with the Futuropolis project, the digital chain was closed and the bridge was made from digital design and construction to digital production. This was a key project for our research program, since it not only demonstrated the architectonically constructive possibilities of computer-aided methods, but also the economic potential that can be opened up by an integrated information technology pipeline, spanning design and production.

项目开始于丹尼尔·李伯斯金于 2005 年为圣加伦大学学生接待处设计的雕塑。这个壮观的作品由 98 个塔组成，意在强调艺术周主题"未来城市"。最初，在建造之前，又一次在 Photoshop 中相对容易地生成了计算机模型。但是随后各个塔间的相关性即使是使用最高端的 CAD 软件也无法绘出。如此复杂的设计，有许多构件组成，其中一个堆叠在另一个里面，无法在给定的时间与规定的预算下搭建起来。最初的搭建预算即使在中国的参与下也还需要大约 80 万

The starting point for this project was a sculpture by Daniel Libeskind that he had designed in 2005 for the student reception week at the University of St. Gallen. This powerful construction, comprised of 98 towers, was designed to underline the artistic theme of the week—'The City of the Future.' What was comparatively easy to design as a computer model—again, in Photoshop—was, at first, as good as unbuildable. The dependencies between the individual towers could not be worked

Futuropolis (St. Gallen, CH)

丹尼尔·李伯斯金为圣加伦大学"未来城市"主题周设计的雕塑透视图。

Rendering of a sculpture by the architect Daniel Libeskind for the University of St. Gallen on the theme of the 'City of the Future.'

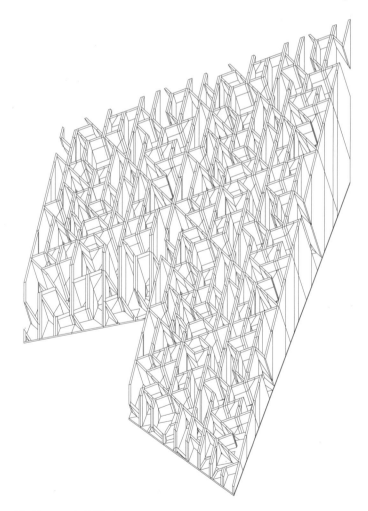

雕塑的 CAD 参数模型。

The parametric CAD model of the sculpture.

瑞士法郎。

在数字链之间的空白没有完全填补时，设计与现实之间存在着鸿沟。就像前面所提到的，如今非常复杂的形式和构造可以在电脑中生成：在过去的十年，必要的CAD软件已成为任何建筑实践的标准工具之一。同样的，在生产领域也有计算机驱动的流程，如CNC（电脑数控加工）机器，它能使单个零件以大批量生产的成本制作出来。这意味着，现今一个铣床与一个普通的复印机没有什么区别。打印一份文档或是切割一个新形状的时候，单独一页纸或是一个MDF零件的成本是一样的。我们所需要的是相应的预先准备好的程序，它可以用来驱动机器。然而，如果数字化设计、数字化建造和数字化生产之间的连接断裂，那么每一个建筑元素的相关数据都要从头开始创建了，这样花费高昂，生产还需暂停，搞得一团糟。

通过缩小数字链的间隙，就可能在可控制的时间和预算内，建造李伯斯金的乌托邦城市。在传统的CAD系统中，我们会先给整个雕塑创建一个参

out, even using high-end CAD software. The complex design, consisting of many pieces stacked one inside the other, could neither be constructed nor completed in the given timeframe and within the stated budget. The first estimates for construction costs—even though they were sourced from China—were around 800,000 Swiss francs.

A gulf opens up between design and reality when the gaps in the digital chain are not fully plugged. As already mentioned, at present very complex forms and constructions can be produced on a computer: in the last decade, the necessary CAD software has become part of the standard toolset of any architectural practice. Also, on the side of production, there are computer-driven processes such as CNC (Computerized Numerical Control) machining which enable single parts to be produced at mass-production costs. This means that, these days, a milling machine is little differentiated from an ordinary printer. A single page or a single piece of MDF is no dearer when a new document is printed out or a new shape is cut. What is needed is the corresponding, pre-prepared construction data that can be used to drive the machine. If, however, the linkage between the digital design, the digital construction details, and the digital production is lacking, then the data for every single building element has to be created from scratch—this is costly, production—halting, and anything but elegant.

By closing these gaps in the digital chain, it became possible to produce Libeskind's utopian city on a manageable timescale and within budget. In a conventional CAD

数化的模型。之后我们除了可以进行必要的结构试验之外，还可以自动生成各个构件精确的几何规格——总计2 164个构件，囊括了必需的辅助构件。这些构件都由计算机来统计，然后用628张木料制作出来。在木料上排列构件时要尽可能地减少产生边角料。Futuropolis现在已准备就绪，360平方米或者说约11.5立方米、7吨重的木料，借助五轴的铣床，在一个集成的数字化程序指挥下全部自动加工而成。建设和生产之间的鸿沟被填平了。

在这个项目中，我们已经表明了数字建造链实现的方法可以从我们实验环境下转换到实际工程中。同样在这个项目的建造上，我们也可从一个新的观点来看待名言"形式追随功能"。不管是在环境中的应用还是从建造过程来看，功能已经失去了它作为形式的前提的地位。从很大程度上来说，如今的建造过程已经从形式中抽离出来。数字链代表着通用的建筑建造公式，它涉及生产过程，包括所有抽象的建筑模块。形式及其设计成为一个摆脱了对功能依赖的变量，现在它能专注于本质的表示方法。

该流程的设计直至最后的细节都展现了其自身优势——以灵活适应性为例：因为重新计算不会花费很长时间，程序中的参数可随

program, we created a parametric model of the entire structure. We were then able to carry out the necessary structural tests, as well as automatically generate the exact geometric specifications for the individual parts—some 2164 in all, including the additional elements that were necessary for the filigree structure. These parts were all numbered by the computer and transferred onto 628 sheets of wood, the layout of the parts arranged to produce as few offcuts as possible. Altogether for Futuropolis, 360m^2 or nearly 11.5m^3 of wood, with a weight totalling some seven metric tons, was completely automatically machined in an integrated digital process, using a five-axis milling machine. The gap between construction and production had been closed.

With this project, we have shown that the methods for the realization of the digital construction chain can be successfully transferred from the environs of our laboratory to real-world projects. And with this project, from the perspective of construction, we can also see the aphorism 'form follows function' in a new light. Function loses its status as a precondition of form, both in the context of usage and from the perspective of the production process. The production process can now be, to a large extent, abstracted from the form. The digital chain represents the general formulation of the building process, its reference process, including all of its abstract building blocks. Form and its design become a variable that—freed from the dependency on function—can now focus on their essence, that is, representation.

This process design has proven itself down to the very last detail—for example, with regards to flexibility: the parameters of a project can be changed at any time, since recalculation does

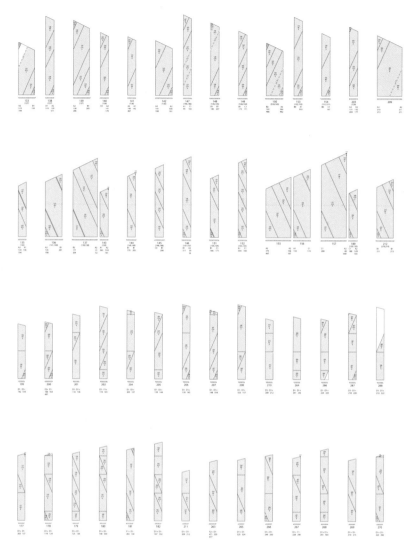

抽取了雕塑中部分构件的列表。没有两个构件是一样的。只有数字技术能够提供快速、无差错的生产和搭建。

An extract from the list of parts for the sculpture. No two parts were alike. Only digital methods were able to deliver fast and error-free production and construction.

切割、铣削、倒角。注：工作站的电脑控制生产流程。

Cutting, milling, chamfering, annotating: workstations for the computer-controlled production.

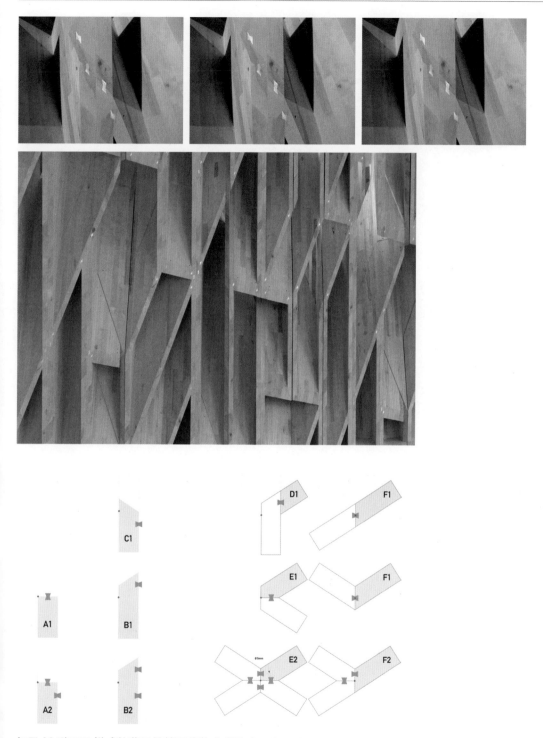

仅需 11 种不同样式的燕尾榫就可将约 3 500 个交叉节点联合起来组成一个整体。

Only eleven different types of dovetail joint were needed to unite the constituent parts through their approximately 3500 intersections

圣加伦大学大厅中完成了的木构雕塑。 The completed wooden sculpture in the hall of the University of St. Gallen.

时更改。在李伯斯金塔的例子中，所有 2 164 块构件的尺寸需重新调整，因为送来的木料是 32 毫米厚，而不是原先预计的 30 毫米厚。这个问题通过简单地改变程序设定中的材料厚度的参数，几小时内就解决了，适应了改变的计算机代码也随之生成。在这样一个过程中，没有什么明显的障碍，因为所有的构件都使用相同的程序进行处理，因此所有的构件不是全部错误——错误一经发现就会立刻得到修正——就是全部正确：机器做事不会半途而废。

最终建成的 98 座塔，高度从 0.2 米到接近 4 米不等，仅用了不到 2 周的时间就手工搭建而成，这多亏了应用于全部构件的自动编号系统和精确的装配指令。每个构件之间的连接由铝制的燕尾榫完成。这种燕尾榫仅用 11 种必要样式就将建造中 3 500 个节点联结成了一体，并且没有使用任何螺丝或胶合物。

李伯斯金复杂的设计作品能够快速的建设、生产、搭建突显了集成数字链的潜力。该项目的成本减少了 70%，其设计建成的最终花费为 11.93 万瑞士法郎。无需人工的编码程序节省了至少 9.2 万法郎；计算机控制的原材料切割又进一步省了约 4.3 万法郎；最后，由于没有通常意义上不可预见或额外损失所产生的意外开支，又节省了约 15.3 万法郎。

not take long. In the case of Libeskind's towers, the dimensions of all 2164 parts needed to be re-adjusted, since the wooden sheets delivered were 32 mm thick instead of 30. This problem was solved within a few hours by simply changing the parameter for the material thickness in the program, and the newly adapted computer code was then generated. The usual snags evident in such a process were practically eliminated, since all parts are dealt with by the same program. Either all parts are wrong—something that is noticed and corrected immediately—or all parts are correct: machines don't do things by halves.

The final construction of the 98 towers, ranging in height from 0.2 m to just under four, was carried out manually and took no longer than two weeks, thanks to the automatically generated numbering system for all the parts and the precise assembly instructions. The connection of the individual elements was carried out using aluminium dovetails, of which only eleven variants were necessary to hold together the 3500 joints required for the construction, using neither screws nor glue.

The comparatively rapid construction, production, and assembly of Libeskind's complex design underlines the potential of an integrated digital chain. The costs were also reduced by 70% on this project: the final cost of bringing the design to fruition was exactly 119,300 Swiss francs. Not having to manually program the code saved more than 92,000 francs; the computer-controlled cutting of the raw boards saved around a further 43,000 francs; and finally, doing without the usual contingency payments for unforeseen, additional incurred costs saved around 153,000 francs.

项　目：	**Swissbau Pavilion**
时　间：	2004—2005
参与者：	Christoph Schindler, Fabian Scheurer, Markus Braach
合作者：	I-Catcher GmbH (Basel, CH) – Felix Knobel, Ruedi Tobler; Fensterfabrik Albisrieden (Zürich, CH) – Peter Eugster; Contec AG (Uetendorf, CH) – Adrian Blain; Bach Heiden AG (Heiden, CH) – Franz Roman Bach, Hansueli Dumelin
参与机构：	Kronospan Schweiz AG (Menznau, CH)

生长的建筑 A Growing Construction

借助 Swissbau 临时展馆，我们全面并令人信服地展示了数字化工具和技术是如何应用于建筑，以及在未来它们将会如何被有效地利用。对于这个工程来说，除了建造和制作，其设计也同样由电脑生成并被整合到数字链中去（尽管对于一个球体来说，谈及"设计"可能有些言过其实了）。

和 Buckminster Fuller 著名的球形馆相比，该工程清晰地展现了一个不受外部或几何上的影响，而是在真实世界背景中由内部自然形成的形式与建构。和章节前面所提及的其他工程相比，决定性的区别在于这个工程并没有先例设计能够提供参考。因此我们不是像在 Futuropolis 里自上而下的方法——从已经完成的设计再到具体的实施细节。Swissbau 临时展馆设计采用了自下而上的方法：最终的形体由单独的细节发展而来，并且是根据预先设定好的规则由电脑设计出来。因为这个原因，Swissbau 临时展馆对 CAAD 所完成的工作来说是一个里程碑。就我们对这个工程的理解，很难再

With the Swissbau Pavilion we demonstrated convincingly and comprehensively how digital tools and technologies can be employed in architecture—and how they will be deployed in the future. For this project, in addition to the construction and production, the design was also generated by a computer and integrated into the digital chain (although, for a sphere, speaking of 'design' is, perhaps, going a bit too far).

In contrast to the famous geodesic domes of Buckminster Fuller, this project shows clearly form and construction not being dictated by external, geometric influences, but rather developing naturally from within, in a real-world context. The decisive difference—with respect to the other projects dealt with thus far in this chapter—is that there was no pre-existing design to work with. So we were not, as in Futuropolis, working with a top-down approach—from the completed design down to the details. The design of the Swissbau Pavilion emerged from a bottom-up approach: The final form grew from the individual details and was designed by the computer according to predefined rules. For this reason, the Swissbau Pavilion is one of the milestones for

Swissbau Pavilion

巴塞尔展示大厅里正在建造中的 Swissbau 临时展馆。

The Swissbau Pavilion during construction in the exhibition hall in Basel.

由矩形构建成的球体的内部和外部实物图。

Interior and exterior view of a sphere constructed from rectangles.

the work carried out in the CAAD department. One could not imagine a more cohesive, more seamless demonstration of the digital chain as we understand it.

The unit cell for the pavilion was an orthogonal, regular cell structure, with twelve surfaces forming the sphere. The software, developed specifically for this project, 'knitted' a network out of a mesh made up of four-sided cells, the unit cells of which could take on different shapes and sizes. The central rule for this process was the definition of the maximum and minimum distance that the four 'knots' of the cell could be from each other: If the distance was too small, the knot would be removed; too large, and the entire cell would be subdivided. In this way, a cellular, completely self-organizing system was generated. An additional complication was the placing of rectangular window openings into the system. We defined these interruptions—as well as the floor openings—to act as attractors for the knots. This allowed us to invert the traditional method of designing such domes. The openings in the sphere would not have to conform to any preexisting structure; rather, the structure would have to adjust itself to the openings, being the windows.

The coffer construction developed, therefore, around the window openings, a digital growth process that was completely programmed and formalized from beginning to end. On the development plans for the pavilion, you can plainly see how flexibly the coffer construction dealt with the awkward 'corners' of the sphere, and how the quadratic base structure of the grid formed, dissolved, and reformed. Free from all geometric rigidity, this pavilion is the first building not to have arisen from the alignment and arrangement of grids, but from a communication

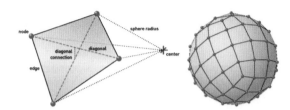

球体有四个角基本部件。通过遵循简单的规则，单元体或扩张或压缩，相互结合，实现自我分布。

The four-cornered base element of the sphere. By following simple rules, the unit cells distributed themselves by expanding or disappeared by coalescing.

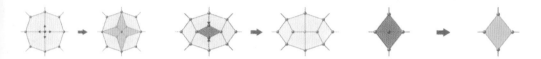

不对称窗户和门的布置借助软件直接转化为变形后的单元体。这确保了那些开口的框架能够和四角的胶合密闭板的边缘相齐平。

The placement of the asymmetrically set windows and doors was directly translated by software into a deformation of the unit cells. This ensured that frames of the openings were flush to edges of the four-cornered plywood coffers.

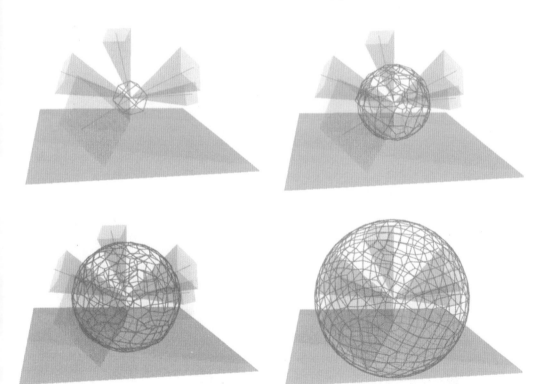

开口在球体表面的投影：由于软件被预先设定，所以建筑自动地根据那些窗户和门做出调整。

The projection of openings onto the surface of the sphere: software is programmed so construction automatically adjusts itself to the windows and the door.

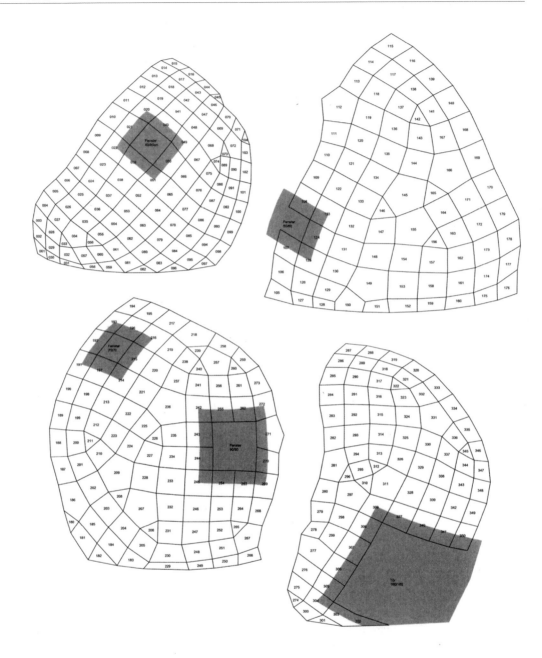

简单的建造说明和清晰的部件编号使得临时展馆能够在几个小时之内被组装起来。

Simple building instructions and a clear numbering of parts enabled the pavilion to be assembled within a few hours.

它是第一个使用四面体网的球体建筑，而这一结构先前在几何学上被认为是无法实现的。但是现在，通过使用这种新的三维排序系统，不可逾越的障碍被轻而易举地克服了。

通过确定衍生出的网格坐标，1 200 片单独建筑部件的精确几何结构被计算出来，并组成了一个高 4 米、半径为 3 米的球体。这包括了对单独部件来说所有必要的连接和钻孔的计算。同样，用来制造单独建筑部件的五轴钻板机的必要驱动也由程序自动生成。

2004 年 Swissbau 临时展馆在巴塞尔会议中心展出。临时展馆的组装仅仅花费了几个小时，包括用人造薄膜为木质结构做覆盖层。覆盖层的样式以及安装 CNC 机器和定制的窗户也由电脑生成。整个工程一共花费了 21 000 瑞士法郎，这比没有应用数字化手段的生产降低了 80% 左右的生产成本。

设计、建造和制作是一个被预先规划好的整体：一个不能被勾画出来的、在灵活性和适应性方面无与伦比的三维体系，这个体系克服了迄今为止不可逾越的障碍。对于我们来说，Swissbau 临时展馆标志着功能、建造和空间的设计机遇，这正由我们通过实践提供给建筑师。

process that remains flexible and versatile and eventually achieves balance. Furthermore, it is the first sphere constructed using a four-sided mesh, something which would have previously been a geometric impossibility. But now, with this new system of spatial ordering, hitherto insurmountable obstacles can be overcome with ease.

From the co-ordinates of the derived mesh, the precise geometry of the 1200 individual construction pieces was calculated for the sphere, with its height of 4m and its radius of 3m. This included the calculation of all the necessary mitering and drilling for the individual parts. Also automatically generated was the necessary program for the control of the five-axis routing machine that was to produce the individual construction elements.

The assembly in the Basel conference center, where the Swissbau Pavilion 2004 was exhibited, took only a few hours, including the cladding of the wooden structure with a synthetic membrane—the pattern for which was also computer-generated—and the installation of the CNC-machined, custom-made windows. The total cost of the whole project was 21,000 Swiss francs, which is around 80% lower than the production costs would have been had the digital chain not been employed.

Design, construction, and production as a programmed unit: a spatial system that cannot be drawn, that is unequalled in its flexibility and adaptability, and which overcomes hitherto insurmountable obstacles. For us, the Swissbau Pavilion is an icon for the extended functional, constructional, and spatial design opportunities that our approach offers the architect.

建造 | Construction

项　目：**Stadsbalkon（荷兰，格罗宁根）**

时　间：2003

参与者：Fabian Scheurer / Participants

合伙人：Kees Christiaanse Architects and Planners (KCAP) (Rotterdam, NL)

赞　助：Kees Christiaanse Architects and Planners (KCAP) (Rotterdam, NL) – Andy Woodcock; ARUP Amsterdam (NL) – Arjan Habraken

圆柱的舞蹈 The Dance of the Columns

就如何应用电脑作为定制工具去完成结构性的任务，并在这一过程中实现结构设计问题的自我处理，"城市露台（Stadsbalkon）"是一个非常好的例子。工程关注一个自行车停放设施，它由 KCAP 建筑师事务所为荷兰城市格罗宁根设计。这个停放设施的顶部被 150 根非常细的圆柱所支撑，而工程的其中一个设计特色就是这些自由排布的圆柱。主要概念之一就是这些圆柱不应该以行列的形式构成一个理想的直角网格，而是应该像森林中的树木一样从地面直接向上生长。这样一个不规则的安排可以描述成一些变量：它们的位置、直径以及每个圆柱的倾斜度。除了整个工程的结构性需要，必须同时保证现存的人行道和自行车道不会被堵塞。考虑到任何一根圆柱的位置都会对整个结构体系产生影响，想简单地用手工方式对圆柱作出安排并不可行——其出发点和北京奥林匹克体育馆类似。同时可能会产生太多悬而未决的问题。假设在第 99 根圆柱被安放好后，有问题突然发生，那么计算工作可能又会需要从头开始做起。

The Stadsbalkon is an especially good example of how a computer can be used as a con figurable software machine for structural tasks, and, in so doing, how problems in structural engineering can almost solve themselves. This project centers on a bicycle parking facility, designed by KCAP architects for the Dutch city of Groningen. The roof of this parking facility is supported by 150 extremely slim columns, and one of the design features of this project is the free ordering of these columns. One of the main concepts was that the columns should not stand in rank and file, artificially forming the points on an ideal, right-angled grid, but rather that they should grow upwards from the ground, like trees in a forest. Such an irregular arrangement can be described with the help of variables: the position, the diameter, and the angle of inclination of each column. Alongside the structural demands of the whole project, it had to be guaranteed that the already existing pedestrian and cycle paths would not be blocked. Since the position of any one column has an influence upon the entire structural system, working out their arrangement by hand under such conditions would simply not be feasible—a comparable starting point to that of the Beijing Olympic Stadium. There would simply be too many balls up in the air at the same time. Should a problem crop up after the 99th column was placed, the calculations would have to start from scratch.

格罗宁根车站自行车库里的支撑物就像丛林中的树木一样被任意地排放。不管是从直径、倾斜角度或者是临近支撑物之间的距离来看，支撑物之间都各不相同。

Supports in the bicycle parking garage for Groningen Station were to be arranged as randomly as trees in a forest. No one support was like the next, neither in diameter, angle of inclination, or distance to its nearest neighbor.

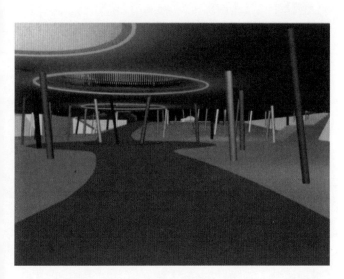

正交结构完全被打散。支撑物和它们的限定参数被预先设定以自动避开那些先前存在的小路。

The grid has completely dissolved. The supports and their parameters were programmed to automatically avoid the pre-existing pathways.

在这个设计中,几何学上的条件(小路、停泊区和顶部的形式)都事先被设定好了。这些限制条件为众多的支撑物界定了"活动区域"。这一系列的图像显示了支撑物在这个区域内是如何扩展直到完全"安定"下来。

In this design, the geometric conditions (pathways, parking spaces, and roof form) were set beforehand. They were defined as a 'living area' for a population of supports. The sequence of images shows how the supports spread out within this area, until they had fully 'settled' it.

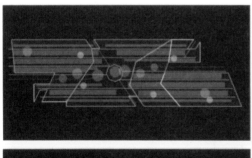
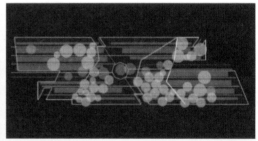
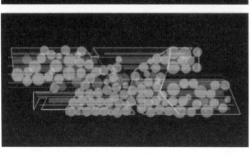
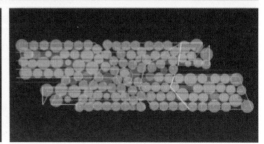

对于支撑群来说,它的适应性条件就是其顶部尽可能使用最少的材料,并确保支撑的位移在一定范围内来承受最大的负重。

Fitness criterion for the population of supports was the greatest load-bearing capacity of the roof using the least material possible and an alternating support offset distance remaining above a certain value.

在计算机中设定任务，避免了这种可能造成的时间和精力上的耗费。我们将系统预先设定为一个动态的微粒模拟体系，在这个体系中圆柱作为独立的媒介，从局部看作用或者是反作用于和它们不直接相邻的圆柱。每一根圆柱都能够移动，自我倾斜并改变自己的直径（从而改变它所能支撑的最大负重）。除此之外，如果圆柱的直径超过了一个特定的值，这根圆柱能够被分成两根。或者相反地，如果直径变得太小，这根圆柱将会完全地消失。其他的限定参数包括基地的边界，现存的必须被清晰地界定为"活动区域"的人行道和自行车道。局部地运作，全局地思考；整个体系的有趣之处就在于它所关系到的屋顶构造的结构完整性。

尽管时间不长，但圆柱的舞蹈是一幅能够移动的壮丽景象。在几个小时之内，计算机就计算出了可行的解决方案，以供建筑师从中挑选出最吸引人和最令人满意的方案。同时因为程序是即时运行的，对于组合中的任意给定点来说圆柱都能够被添加或减少。

虽然结构设计师和建筑师之间的潜在冲突在这个设计阶段经常会突然出现，但是也因此能够在虚拟层面得到解决。在未来，凭借着特定的结构和设计能力，计算机甚至能够脱离现实世界中对结构难题的处理方法，在科技和建筑之间起到一个很有意思的协调作用。

By programming the task into the computer, this time and effort was avoided. We programmed the system as a dynamic particle simulation, with independent agents—the columns—that, when viewed locally, would either act or interact, as appropriate, in their own interests against the columns that were not in their immediate vicinity. Each column could move, incline itself, and change its diameter (and with it, the maximum load it could support). On top of that, a single column could either split into two if its diameter exceeded a certain value or, conversely, disappear altogether if its diameter became too small. Further parameters, such as the site boundaries and the existing pedestrian and bicycle paths that had to be kept clear defined the 'playing field.' Act local, think global; the interests of the overall system was what counted: the structural integrity of the roof construction.

The dance of the columns was a moving spectacle, albeit short-lived. Within a few hours, the computer had calculated the possible solutions from which the architects could choose the most attractive and best suited. And since the program ran in real time, columns could be either added or subtracted to the mix at any given point.

Possible conflicts between structural engineer and architect—which always crop up in this phase of the design process—are therefore already solved at a virtual level. With its specific structural and design capabilities, the computer could, in the future, take on an interesting mediating function between technology and architecture that would even go beyond the solution of real-world structural problems.

项　目：**Metrostation**（意大利，那不勒斯）
时　间：2005 Period
参与者：Fabian Scheurer
合作者：B+G Ingenieure, Bollinger und Grohmann GmbH (Frankfurt a. M., D) – Arne Hofmann

稳定性的颠覆 Foreshaking Stability

这个工程与法兰克福的 Bollinger+Grohmann 工程师事务所合作完成。项目不仅仅在工程层面非常有趣，同时它也为我们基础研究中的一些非常重要的方面指明了方向。

项目起源于那不勒斯地下车站入口处的顶棚。该顶棚由五个不规则的顶板组成，各部分分别占据 250 平方米的面积，包括 800 个钢铁元件和 289 个节点。结构上的稳定性和顶部的三维属性完全是以折叠的方式或是节点之间不规则纵向位移形成的。但是这些位移必须限制在先前界定的范围之内。问题是，不太可能去人为地开发一套能遵循最初构想且能生成稳定的不规则结构系统体系。如果人为去做，这样一个复杂的建筑就很难被简单地分析清楚。

所以我们把正常的分析次序反过来，仅仅关注于结果，从最后的结果开始计算。换句话说，为了能够利用这些数据一步步得到最优解，我们模拟其中一个顶棚部分的结构作用。简言之，

This project came about in cooperation with the Frankfurt-based engineering practice of Bollinger+Grohmann. It is not only the scale of the project that makes it interesting, but also the fact that it casts light on certain aspects that are important for the fundamental approach to our work.

The starting point for this project was the design of an entrance canopy for an underground station in Naples, made from five irregularly arranged roof segments, each $250m^2$ in area. Each of the five roof segments was made up of 800 steel elements with 289 nodes. The structural stability and the three-dimensional nature of the roof were to be provided solely by the way it was folded, or rather by the differing and irregular vertical displacements of the nodes. These displacements would, however, have to stay within certain predefined boundaries. The problem was that it was impossible to manually develop any meaningful system of rules that would lead to a stable, asymmetrical construction that would conform to the original idea. Manually, such a complex construction could simply not be analyzed.

So we turned the normal sequence of analysis around, and started calculating backwards, keeping our eyes only on the result. In other words, we simulated the structural behavior of an entire single roof

利用合适的软件生成并分析完整的顶棚结构是可行的。通过使用进化算法，这个结构自身得到了优化，展现在我们面前的是一个类似北京奥林匹克体育馆项目的简化形体。这个计算程序的"基因组"包括了 z 坐标轴的所有 298 个节点，并且通过这些节点生成了 40 个随机的顶棚。建筑的适应性标准就是每个节点产生的最小垂直向下方向的可能位移。每一代产生 40 个变体，200 代之后，这个结构得到了明显地改进：没有一个节点向下的位移超过 126 毫米——考虑到 50 米的大跨度，这是一个非常重要的结果。同样，不可能用手工的方式来计算得出这个结果。

在我们看来，这个工程有两个方面尤其值得关注：第一，事后很难去理解为什么通过计算机能得到这样一种几何学结构。我们甚至很难说这就是所有可行的处理手段中最好的一个。你只能说这是一个非常好的功能性处理手段。更笼统地说，唯一有价值的就是最终的结果——一个稳定的结构而且无法显性地解析其稳定的原因。

第二点值得关注的就是经过计算最后生成的顶棚的实际形式。最初，有一套规则体系以及预先定义的特征，但是却没有预先确定设计。然而在这个案例中，尽可能地接近最初的设计却是其中的一个质量标

egment, in order to be able to optimize it step-by-step by using this data. More succinctly: With the appropriate software, it was possible to generate and analyze a model of the complete roof structure. The structure itself was optimized using an evolutionary algorithm, a simpler form of which we saw in the project for the Beijing Olympics Stadium. The 'genome' for this algorithm contained the z-coordinates of all 298 nodes and, from these, 40 random roofs were generated. The criterion for the fitness of the construction was the smallest possible downward vertical displacement of each node. After 200 subsequent generations of this genome in 40 variants each, the structure had been significantly improved: None of the nodes moved downwards further than 126mm—a significant result, considering the large span of 50m. And, of course, this could never have been calculated manually.

In our opinion, two aspects of this project are particularly worthy of attention: First, it is very difficult after the event to understand why the geometrical form arrived at by the computer looks the way it does. One cannot even say that it is the best of all possible solutions. The only thing that you can say is that it is a very good, functioning solution. More generally put: The only thing that counts is the end result—a stable structure—and the reasons for this stability cannot be explicitly formulated.

The second point worthy of attention has to do with the actual form of the roof produced at the end of the computational process. At the beginning, there is some basic predefined quality, a set of rules. There is no predetermined design. In this case, however, the proximity to

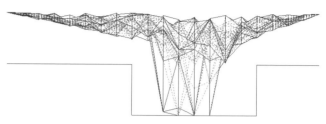

该树状的支撑结构是由建筑师 Dominique Perrault 为那不勒斯的地铁站所设计，如果单纯使用人工的方法是不可能实现其结构稳定。

This tree-like supporting structure by the architect Dominique Perrault for a metrostation in Naples could not be stabilized using manual methods.

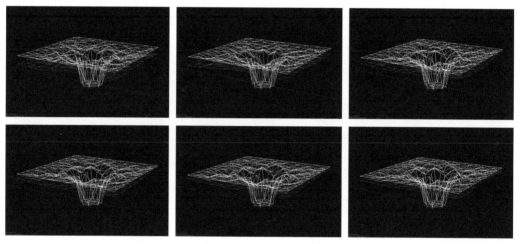

这一系列的图显示了在使用了遗传算法后树状支撑结构在稳定性方面的改善。在最初的设计中（左上），位于节点—柱梁支撑结构（用白色显示）边缘的原件在负重情况下出现了不能接受的倾斜（用黄色显示）。结果（右下）显示，仅仅通过最小程度的节点的位移（换句话说，几乎不改变设计），减少了50米跨度的倾斜，达到了能够接受的范围。

The sequence shows the improving stability of a tree-like supporting structure using a genetic algorithm. In the original design (top left), elements at the edge of the node-column supporting structure (shown in white) deflected unacceptably under load (shown in yellow). The result (below right) shows that with only a minimum displacement of the nodes (in other words, by hardly changing the design) deflection over a span of 50 m was lowered to acceptable limits.

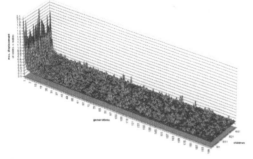

节点和柱梁的支撑结构通过一个遗传算法得到了改善。z 轴展示了不同节点的位移；y 轴显示了负重；x 轴显示了在不同负重情况下节点的位移得到的改善。

The node and column support structure is improved by a genetic algorithm. The z-axis demonstrates the displacement of various nodes, y-axis, under load. The x-axis shows that displacement of the nodes under load improves from generation to generation.

准——事实上我们实际生成的设计只和最初的设计相差了几米。这同样显示了当从头开始建造时,科技具有非凡的适应性。不管怎么说,产生的结果都有更大的实用性和影响力。地下车站的顶部看上去和建筑师最初的设计没有什么不同,但是它在结构上更稳定——这也就是两者最大的差别所在。

和北京奥林匹克体育馆或 Swissbau 临时展馆的例子相比,考虑到计算所需要的经费过于庞大,这个难题解决方案无法即时生成——至少利用当今的计算机科技难以实现。40 个变体的 200 代算法提供了 8 000 种不同的处理手段,它们中的每一个都需要大约三四秒钟的估算时间。对于 5 个顶棚部分中的每一个部分来说,总的估算时间大约需要 7 小时,甚至是一个晚上的时间,所以我们能够推算出一个完整的顶部大概需要 5 倍的时间。但是和手工的处理手段相比,仍然算得上是闪电般的速度(手工的方法,无论从意向上还是从目的上来看都是不可行的)。

随着电脑硬件速度的飞速提高,我们不应该低估在这些机器上运行的算法的质量。在协调选定目标之间的平衡的过程中,这些计算程序起到了至关重要的作用。

the original design was one of the quality criteria—and the design that we produced actually only differed by a few centimeters from the original. This also shows just how adaptable the technology can be when it is properly set up from the start. At any rate, the result produced has greater functionality and impact. The roof of the underground station looks no different from the architect's original design; it is, however, structurally stable—and this is where the great difference lies.

In contrast to the examples of the Beijing Olympics Stadium or the Swissbau Pavilion, the solution to this problem can not be generated in real time—at least not with the currently available computer technology—since the calculation overheads are simply too great. Two hundred generations with 40 variants each produces 8000 individual solutions, each of which needs around three to four seconds of computation. This gives a total calculation time of around seven hours—or overnight—for each of the five roof segments, and we can extrapolate this to around five times longer for the complete roof. But this is still lightning fast compared to a solution by hand (which would be, for all intents and purposes, impossible).

While the speed of the computer hardware may be growing ever faster, we should never underestimate the quality of the algorithms run on these machines, which plays just as central a role in such complex tasks as the corresponding weighting of the chosen criteria.

项 目：Monte Rosa（瑞士）

时 间：2006—2008

参与者：Philipp Dohmen, Markus Braach, Kai Rüdenauer, Benjamin Dillenburger, David Sekanina, Christoph Schindler

合作者：Abteilung Bautechnologien EMPA (CH) – Zimmermann Mark;
Häring Holz- und Systembau AG (Pratteln, CH) – Armin Röhm;
Häring Fenster und Fassaden AG (Frenkendorf, CH) – Lothar Müller

时 期：Trumpf Maschinen AG (Baar, CH); Hundegger Maschinenbau GmbH (Hawangen, D); Thyssen Krupp Nirosta GmbH (Krefeld, D)

数字小屋 The Digital Chalet

从某种程度上说，该项目是以前所有项目技术的统一，甚至是各种以前存在的解决建筑问题手段的统一。新建的 Monte Rosa 小屋位于瑞士最高峰海拔 4 634 米的杜富尔峰的山脚。它基于苏黎世联邦理工大学的一个学生设计，指导教师是 Andrea Deplazes 教授。以它最原始的形式，该设计原本既不经济也没有技术的可行性。海拔 3 000 米的建筑需要特殊的条件：首先是每种建筑构件的重量，这些都不得不通过直升机运输；其次是整个工程的结构，它将要承受高风速、雪荷载和温度上的急剧变化，同时还有一些其他不利的室外条件。最后，必须考虑到场地上可用的时间必然受天气条件的限制。在这种条件下建造雄伟和复杂的建筑是一项挑战。然而，我们希望展示通过计算机辅助设计、建造和生产方式可以应对这种挑战。更简单地说：只有在数字化产业链完全做好准备的时候，像 Monte Rosa 的设计才有可能实现。

To some extent, this project is a consolidation of all the previous projects, technologies, and, above all, the various approaches to architectural problem-solving presented previously. The new Monte Rosa chalet at the foot of the Dufour peak—at 4634m, the highest in Switzerland—is based on a student design from the ETH, produced under the tutelage of Professor Andrea Deplazes. In its original form, the design was neither economically nor technologically feasible. Building at 3000m above sea level requires special rules: first, the mass of the individual building elements, which have to be transported by helicopter; then, the structural solution of the entire project, which will have to withstand high wind speeds, snow loads, and extreme differences in temperature—as well as many other adverse outdoor conditions. Finally, the amount of available time on site—necessarily limited by weather conditions—must be considered. Producing ambitious and sophisticated architecture under such conditions is a challenge. We hope to show, however, that this challenge can be met by employing computer-aided design, construction, and production methods. More succinctly put: A design like Monte Rosa can only be achieved when the digital chain is fully primed.

当我们生成第一个把所有参数——从墙厚到床的数量——都考虑到的电脑模型时，我们碰到一个相对实际的问题：与其他山地小屋相比，原始设计里的每平方米内床的数量太少，必须有所提升。在第一章讨论的将一块建筑用地分成若干土地的过程同样适用于山区小屋的床的布置——它们是自组织的。我们根据一个动态的可以四处移动的墙体和表皮来获得优化布置的模型，在一定的框架里让电脑处理房间尺寸和可能的家具布置。例如，它不能干扰设计的某些关键几何要素，比如 80 平方米倾斜 10° 的太阳能立面。我们基于这样的事实：表面积为 10 平方米的墙体如果尺寸增加 1 平方米，它的成本不会自动增加 10%。我们首次整合了一个建筑耗费和建筑面积或者建筑部件尺寸之间的一个非线性联系并进行计算。这些让我们能估计整个项目的成本。每平方米床的数量、地板的面积、价格、尺寸或者立面面积这样的依赖关系可以实时地，直接地被采用。用这种方式，设计的体量和底层平面被完善成数字链的第一个步骤。Monte Rosa 小屋的总体几何结构被当做一个建筑系统转化为程序化的 XML 文件。

When we generated the first computer models that took all parameters into account—from the wall thickness through to the number of beds—we hit a relatively mundane problem: The number of beds per square meter in the original design was too small compared to the other mountain chalets, and needed to be raised. The process of dividing a building plot in parcels of land, discussed in the first chapter, also holds true for the beds in mountain chalets—they are self-organizing. We let the computer handle the size of the rooms and their possible furnishing according to a dynamic model that could move walls and facades around in order to obtain an optimal layout, albeit within a certain framework. For example, it was not to disturb certain key geometrical aspects of the design, such as the solar facade inclined at 10° with its area of 80m^2. For the first time, we integrated a non-linear correlation between building costs and either building area or building component size into the calculations, based on the fact that a wall with a surface area of 10m^2 will not automatically be 10% more expensive if increased in size by one square meter. This allowed us to generate a differentiated cost appraisal for the entire project. Dependencies such as number of beds per square meter, floor area, price, size, or facade area were directly assessable, in real time. In this way, the volume and the ground plan of the design were refined as the first stage in the digital chain. The overall geometry of the Monte Rosa chalet was transferred as a building system to a programme-independent XML file.

Construction

新 Monte Rosa 小屋位于海平面 2 883 米的 Untere Plattje 花岗岩上，被 Gorner、Grenz 和 Monte Rosa 三个冰川环绕。

The new Monte Rosa chalet is located on the 'Untere Plattje' granite promontory 2883 m above sea level, surrounded by the Gorner, Grenz, and the Monte Rosa glaciers.

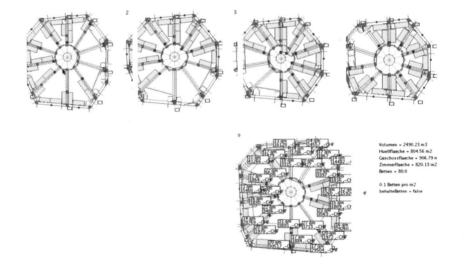

小屋的参数化建筑模型能够迅速而且准确无误地进行复杂几何体的分析。

A parametric building model of the chalet enables a quick and error-free analysis of the complex geometry.

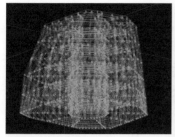 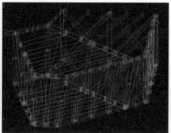 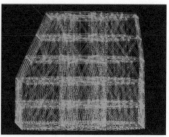

承重结构被建成参数化模型,它的重量通过进化算法进行最小化。

The load-bearing structure was parametrically modelled and its weight minimized using an evolutio-nary algorithm.

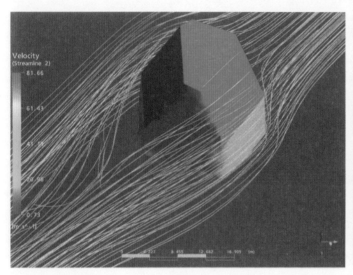  

风荷载的模拟在 Monte Rosa 小屋的形体塑造中起到了重要的作用。

Simulations of wind loads played a significant role in shaping the form of the Monte Rosa chalet.

在生产大厅里对一个试验单元进行数字化预切割的每个部分。

The digitally pre-cut individual elements of a test module in the production hall.

整个建造过程和每个建筑元素的性能在如 Monte Rosa 小屋这样的建筑中扮演了特殊的角色。因为和更适宜的环境相比,在如此高的海拔位置上,强度与重量的比值更是一个决定性因素。为了模拟预期的荷载,建筑的形式采用了风洞试验,然后按照获得的荷载结果进行深化发展。因此,我们使用了原本用于汽车工业开发中作为碰撞模拟的程序,这使我们能够估计结构的受力和移动情况以及不同材料的特定表现。

开发最轻且最稳定结构的任务导致了一个结果,从结构的角度看,半木结构的房子的建筑构造比普通的线性承受荷载的结构更有效率。另外,整个结构的强度重量比被进一步地完善和改进,我们削减了每个元素以获得可能的最高值。这种优化使用一种进化算法来实施,该算法通过一个任意数量的迭代运算,试图在规定的框架参数中获得一个优化的方案。因而我们能够减少整个建筑物重量的 40%——每个建筑元素都将要通过直升机来运输,因此这是一个相当大的节省。

The structural performance of the entire construction and of each individual building element plays a special role for a building like the Monte Rosa chalet, since strength-to-weight ratio plays a much more decisive role in such an elevated position than it does in more moderate environments. To simulate the expected loads, the building form was subjected to a flow test, and was developed further, in line with the loading results obtained. For this purpose, we used programs that had been developed for crash simulations in the automobile industry; these allowed us to evaluate the stresses and movements of the structure or the specific behavior of various materials.

The task of developing the lightest and most stable structure possible led to the result that—from a structural point of view—a building construction resembling a half-timbered house was more efficient than the normal linear load-bearing structures. In addition, the entire structure's strength-to-weight ratio was further refined and improved; every element was pared down to obtain the highest possible value for this relationship. This optimization was carried out using an evolutionary algorithm that worked through an arbitrary number of iterations in order to reach an optimized solution within the defined framework param-eters. Thus, we were able to reduce the weight of the entire construction by 40%— a considerable saving, in view of the fact that every single building element would have to be lifted by helicopter.

整个过程的前提条件是，用于组成数字链的软件接口是可扩展的。下一步用于建造木结构建筑时，这点显得尤为重要。由于每块构件的非标准性质，轻型木结构房屋的建造需要花费更多的时间。这个事实给工程带来了额外的复杂性，它可以通过自动生产过程来解决。在早期的工程中我们已经显示了，木构连接能够以任何可以想象的形式制造出来，不管一次制造过程需要多少个构件，都是精确而且便宜的。然而，这样做的话就需要一个集成化的建造和生产过程。Monte Rosa 小屋的木结构的制造时间实际上不足60小时。立面无缝地整合到这样的工序中：由两块磨砂不锈钢板制成的真空隔绝的嵌板不仅能够满足功能和设计的所有要求，还能够通过电脑控制的机器人来进行精确到毫米的切割和焊接。

这个工程的成果是一个融合许多数字工具的集成系统，从规划到建造和制造。Monte Rosa 小屋表明，如果你利用了现代信息技术工具，在不寻常的地点建造不寻常的建筑也不会带来额外的预算。

A precondition for the entire process was that the interfaces of the software building the digital chain were permeable. This was even more true for the next step of producing the wooden construction. Balloon-frame buildings are a lot more time-consuming to build, thanks to the non-standard nature of their individual pieces. While this fact brings additional complexity to the project, it can be managed by the application of automated production processes. As we have already shown in earlier projects, wood connectors can be produced in almost any form imaginable, precisely and cheaply, no matter how many are required from a single production run. However, to do this, an integrated construction and production process is required. The schedule for the demanding wooden construction of the Monte Rosa chalet was actually produced in less than 60 hours. The facade was integrated seamlessly into this process: Vacuum isolation panels, made from two sheets of stainless steel with a granular filling, were able to fulfill not only every demand as regards function and design, but could also be cut and welded to millimeter accuracy using computer-controlled robots.

The achievement of this project is the amalgamation of many digital tools into an integrated system, from planning through construction and on to production. The Monte Rosa chalet demonstrates what possibilities emerge if you take advantage of the tools offered by modern information technology, when building extraordinary architecture in extra-ordinary places—but without extra-ordinary budgets.

第五章

过程和平衡

从工业化社会向信息化社会的转变伴随着新的思考方式、新的操作方式以及新的世界观。塑造了工业社会机械论世界观的基础操作原则（虽然目前或多或少存在一些残留）已经被一个基于交互循环和网络的原则取代了。

尽管在本书中"过程"一词已经不止一次用来描述前瞻性的世界观和创新性的生产方法，它已经成为一个理解工业化初期阶段的关键。简而言之，超过150年的手工生产方式被对能源和材料流的控制所取代。所以，正如在第一章（一个新的深层结构）中提及的水晶宫所表现出来的一样，这些过程对于建筑来说同样变得日益重要。

然而，"过程"一词的意义和应用在工业社会和信息技术社会是明显不同的。一个可行的过程理论才出现了几年，而软件技术起到了核心作用，正如SAP的例子所表明。即使在1970年代末，"过程"一词在管理学的课本中也鲜为所见。过程的实施也有重大变化，它们不再是单向的，而是双向的。对一个流程的严格控制不再至关重要，也不是一个合乎逻辑的基本运行准则。相反，基本的工作原理的重点是利益平衡和生产均衡（同时在个体之间以及系统本身内部）。过程的控制不再像是生硬而快速的总体规划，在一个特定的地点和特定的时间框架内确定每一个步骤。现在，一个完全平衡的系统是在软环境和宽泛的框架中产生的。更简洁地说：如果一个塔在一个特定地点需要加高，那么在另一个地点的建筑物将要变低，成本也会有所变化。这样，整个城市规划设计理念保持了稳定的平衡。然而在这种新形势下，稳定绝不能混同于理想化的和谐。相反，稳定是指一个系统中的元素处于健康竞争中，因此它们一直保持"健康"，足以适应任何条件的变化。

当然，这种方法并不是全新的：第一个全面的互惠主义理论，如热力学和现代进化理论出现于19世纪，其次在20世纪中叶出现了控制论和系统论。在科学领域，在相对论和量子理论以及生物学、化学或经济学中，关系的概念早已取代物质的概念。

历史就说这么多了。当今世界的新特征在于——多亏了现代信息和通信技术——我们第一次有了可用的工具，来制定和实现例如互惠系统这样的平衡。目前，我们正在经历新奇的理论，思维方式正成为我们的技术构建环境的重要组成部分。技术的成功取代了来自于老式的、低效范式的压力。同时，我们正从各种危机中学习——从恐怖的人为气候变化到能源危机到全球金融系统振荡，我们无从得知新技术将会如何与旧范式结合。看来，我们很可能也很快要学会放弃明确的控制，抛弃普遍的参考框架（例如网格），包括它们最小、最细微的细节。如果我们想要让这个世界容纳多样性的个人利益与项目，那么我们要学习很多东西。

Process and Balance

The transition from an industrial to an information technology-based society is marked by a new way of thinking, a new way of operating, and a new understanding of the world. The grounding operational principle that shaped the mechanistic world view of the industrial society—the relics of which are still more or less active—has been displaced by a principle based upon interdependent cycles and networks.

Although the term 'process' has been used more than once in this book to describe pioneering world views and innovative methods of production, it has already been a key term for understanding the early phases of industrialization. Simply put, over a period of 150 years artisanal means of production were replaced by the control of the flow of energy and materials. Therefore, these processes also became increasingly important to architecture, as shown by the example of the Crystal Palace, mentioned in text I (A New Deep Structure → pp. 1 ff.).

However, the meaning and application of the term 'process' differs markedly between the industrial society and the information technology society. A feasible theory of processes has only been around for a few years and, here again, software technology has played a central role, as illustrated by the example of SAP. Even at the end of the 1970s, the term 'process' was almost non-existent in textbooks on management. The implementation of processes has also changed significantly; they are no longer unidirectional, they are bidirectional. The strict control of the flow of a process is no longer crucial, neither is a logical, base operational principle; rather, it is about the balancing of interests and producing equilibrium, both between individual elements and within the system itself. The control of processes is no longer achieved through some hard-and-fast master plan that fixes every step in a certain place and within a certain time frame. Now, it is achieved in the context of soft and only broadly framing constraints, within which a totally balanced system is produced. More succinctly put: If a tower needs to be built higher on one particular site, then on another site the buildings will have to be lower and more cost-conscious. In this way, the entire urban planning concept is maintained in stable equilibrium. However, in this new context, stability must never be confused with idealized harmony. Rather, stability means that the elements of a system are in healthy competition, so they remain 'fit' enough to adapt to any changing conditions.

Of course, this approach is not a new one: The first comprehensive theories of reciprocity, such as thermodynamics and modern evolutionary theory, arose in the 19th century, followed by cybernetics and systems theory in the mid 20th century. In science, the concept of relationship has long replaced that of substance, in relativity and quantum theory as well as in biology, chemistry, or economics.

So much for history. The novelty in today's events lies in the fact that—thanks to modern information and communications technology—we have, for the first time, the tools available which enable us to formulate and implement such equilibria, such reciprocal systems. Currently, we are experiencing that

novel theories and ways of thinking are becoming important components of our technologically constructed environment. Technological success places enormous pressure on old, less capable paradigms. Simultaneously, we are learning from various 'crises'—from the feared anthropogenic climate change, to the energy crisis, to the quakes in global financial systems—how the coupling of new technological possibilities with old paradigms is leaving us out of our depth.
It seems that we very probably—and soon—need to learn to give up explicit controls, to let go of universal frames of reference (like, for example, the grid) purporting their extension down to the smallest or even microscopic details.
We have much to learn if our planet is not to become too narrow for our many and varied individual interests and projects.

Charles Sanders Peirce:
Lectures on Pragmatism

' Consider what effects, that might conceivably have practical bearings,
we conceive the object of our conception to have.
Then, our conception of these effects is the whole of our conception of the object.
The idea of a thing consists simply of the practices that imply it.
In order to develop the meaning of an idea,
one must simply define the practices that generated it. '

结构

Structures

项　目：Pavillons

时　间：2001—2002

参与者：Markus Braach, Oliver Fritz, Christoph Schindler, Odilo Schoch

学　生：Philipp Dohmen, Jenny Donno, Ulrike Horn, Johann Käding, Rüdiger Karzel, Nils Kemper, Oskar Zieta

数字建筑结构 Digital Building Construction

很长一段时间以来，钢板和它的功能都是我们在 CAAD 实验室中的研究项目重点之一。对于钢板的思考和工作给我们提出了特别的挑战。在规划构建几何形状时它要求极度精确和细致，目前很多建筑院校广泛地使用聚苯乙烯块，而且大多数都是用多轴切割机来加工塑形。利用这种方法，可以将一整块材料雕刻出形体。在另一方面，钢板是一种在处理时需要仔细地建造规划的材料。不像木材或者混凝土，钢板不能容许构想、建造或者生产中的错误：它是一种要求极其严格的材料。同时它把建筑教育与高度自动化的钢板加工技术联系在一起。这种科技经常用于汽车工业并且开拓了有意义的前景，在很大程度上沿袭了工业化建造生产先驱（如 Jean Prouvé）的传统，但如今很少使用了。

For a long time, sheet steel and its function has been one of the focuses of our research program in CAAD. Thinking and working in sheet steel presents particular challenges. It demands extreme precision and meticulousness in its planning and construction geometry, in contrast to working with polystyrene blocks, which are so popular in architecture schools and are mostly shaped with a multi-axis milling machine. Using this method, sculptural forms can be produced from a single block of material. Sheet steel, on the other hand, is a material whose handling needs careful constructive planning. Unlike wood or concrete, sheet steel forgives no errors in conception, construction, or manufacture: it is an extremely difficult and demanding material. It also links architectural education to the highly automated sheet steel-working technology, so often used in the automobile industry and thus opens up interesting perspectives, very much in the tradition of the pioneers of industrialized building production like, for example, Jean Prouvé—a rare approach nowadays.

因此，我们对两个相关的问题产生了兴趣：第一个关于材料，它的各不相同的特征，它不同的工序能力，以及不同的节点技术。如何把一种平整的、不稳定的、不均匀的材料转变成一个稳定的、精确的三维形式？此外如何从中发展出定

Consequently, we are interested in two related questions: the first regarding the material, its varied characteristics, its many processing capabilities, and the different jointing technologies. How might a flat, unstable, and astoundingly non-homogeneous material be turned into a stable and precise three-dimensional form, and

已经切割好的 NDS 展馆的钢板单独构件正等待运输。

The ready-cut sheet steel individual elements of the NDS Pavilion awaiting transportation.

只有在组装的时候才需要人力。

Manual labor was necessary only for the assembly of the Pavilion.

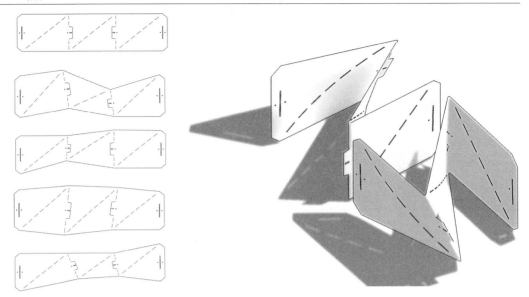

每个单独元件开始都按二维元素在电脑上绘制，然后折叠起来展现出最后的三维形态。

Each individual piece was firstly drawn as a two-dimensional element on the computer and only afterwards folded to take up its final three-dimensional shape.

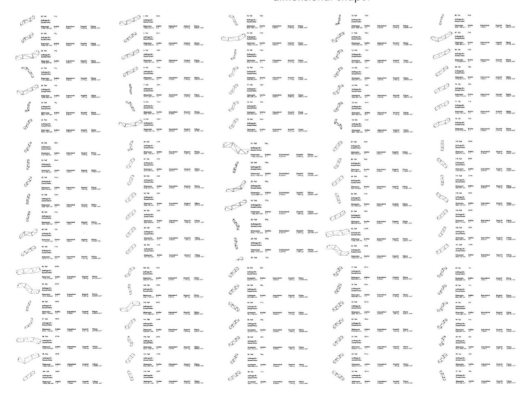

一张二维构件的列表和其源代码的细节。

Detail of a list of the two-dimensional segments for the construction and its source code.

组装完成的第一个展馆的细节。

Detail of the first pavilion after its assembly.

制的系统？另外一个问题是关于对已经长期使用于金属处理工业中的复杂机器的理解，这种机器展示出了非凡的灵活性和多功能性。磨碎和弯折，挤压和切割——几乎没有其他的材料可以用如此多的方式处理。钢板的实验是数字建筑建造新形式的第一个例子。大规模生产的产品结合——比如螺丝钉、断面和模具——不再是一项建筑工程中的首要元素，反而，通过将抽象的过程更进一步，整合构架上最为相似、甚至等同的电脑控制机器制造过程才是最重要的元素。取代整合许多不同的大规模制造的零件来完成各种各样的想法，建造现在是由少量几何的、结构性的、专门为建造定做并以单一或类似材料构成的构件组成的。

我们实验室的首座大型钢板项目之一是 SXM 展馆，于 2002 年夏天设计并落成。也许这个建筑没有满足作为展馆所有的功能要求，但作为一个在设计、建造和生产中应用信息技术的宣言，它是一个令人瞩目的成功。416 块零件每个都不一样，而且使用了一个集成的数字流程，每一块都在一个机器上经过一个单独的顺序完成建造、定形、变形过程，最后被生产出来。

建筑的概念来自于中心交叉紧密连锁在一起的 N 或 Z 形状的元素——这是建筑的一个重要的特色。也许并不是很直观，利用这个 N 和 Z 的系统，能设计出许多不规则的形式：沿着平坦的表

furthermore, how might customized systems be developed from it? The other question concerns an understanding of the complex machines that have long been used in the metalworking industry, which display outstanding flexibility and versatility. Milling and bending, pressing and cutting—almost no other material can be worked in such a variety of ways. The sheet steel experiments were the first examples of a new form of digital building construction. No longer is the combination of mass-produced products—like screws, profiles, and mouldings—the primary element in a construction project: Rather, taking the process of abstraction one step further, it is the combination of the most structurally equivalent, or even constant, computer-controlled machining processing. Instead of combining many different mass-produced parts with various implementations in mind, a construction is now composed of a few geometrically and constructionally made-to-measure components from a single, homogeneous material.

One of the first big sheet steel projects in our department was the SXM Pavilion, designed and produced in summer 2002. The construction may not have fulfilled every requisite for its function as a pavilion due to its design, but as a manifesto for applied information technology in design, construction, and production, it was a resounding success. Each one of the 416 pieces was unique and, using an integrated digital process, each piece was constructed, configured, transformed, and finally produced on a machine in a single operational sequence.

The concept for the construction was based upon N- or Z-shaped elements that interlocked around a central intersection—a key feature of the construction. Perhaps counter-intuitively, with this system of Ns and Zs a variety of irregular forms could be designed: Sheets folded through two or three dimensions could be produced alongside

面通过二维或三维的弯折被生产出来。独立元素的几何细节由电脑生成，基于一个从 N 和 Z 剖面出发的基础模块，将三维坐标变化或折叠到一个二维平面上，同时包括所有必要的穿孔、开洞以及为了后来组装而作的标记。反过来，收集的数据能被直接导入激光切割机的控制软件。无论是只用一次的还是一系列相同部件中的一个，机器生产每一块构件的时间都不到 1 分钟。有关成本降低标准化的问题——例如 SXM 展馆的两或三个元素是否应该是同样的形式——实际上与建造和生产不相干。这个过程类似于一个普通的激光打印机——打印一页文字十次或一次打印十页的文字都没有什么区别。即便如此，所有的零件均按照一样的原则建造，确保了稳定的、环环相扣的节点一起组成微妙和几何上灵活的建筑。当把 N 和 Z 形的零件折叠在机器里，并且电脑自动给这些组装零件编号，这样展馆的建造就没有什么困难了。

相互紧锁的节点和整个结构的关系也是这个七天临时展馆的一个主要特点，它是在 2002—2003 年度的冬季学期发展出来的。相比 SXM 展馆，其结构复杂程度较低。然而，如它的工作题目暗示的，它在一周时间内完成设计、编程、制造和组装。

这两个展馆迈出了第一个试探性步伐——使用难以加工的材料和要求苛刻的机器。在通往一个新的、数字建筑制造方式的途中，一条充满更多惊喜的路还在等着我们。

flat surfaces. The geometrical details of the individual elements were generated by computer, starting with a base module built from an N and a Z section, along with the transformation or folding of the three-dimensional coordinates onto a two-dimensional plane and including all the necessary perforations, holes and markings for later assembly. In turn, the data collected could be imported directly into the control software for the laser cutting machine. Each piece could be produced by this machine in little under a minute, whether it was a one-off piece or one of a series of identical pieces. Questions about cost-lowering standardization—whether, for example, two or three elements of the SXM Pavilion should have the same form—were thus irrelevant for both the construction and the production. The process is analogous to an ordinary laser printer—it makes no difference whether it prints one page of text ten times or ten pages of text once. Even so, all pieces were built according to the same construction principle, ensuring that stable, interlocking joints held together the delicate and geometrically flexible construction. After folding the Ns and Zs in the machine press, the computer-generated assembly markings made for a problem-free build of the pavilion.

The relationship between the interlocking junctions and the entire structure was also a key feature for the seven-day Pavilion that was developed over the winter semester of 2002/2003. In comparison to the SXM Pavilion, its structure was less complex. However, it was—as its working title suggests—designed, programmed, produced, and assembled all within a week.

These first two pavilions were the first tentative steps—with difficult-to-work material and demanding machines—on the way to a new, digital building construction method, a path along which more surprises yet await us.

项 目：**Paravents**
时 间：2006—2007
参与者：Oskar Zieta
合作者：Trumpf Maschinen AG (Baar, CH)
mmenarbeit / Cooperation

学习用机器建造 Learning to Build with Machines

我们部门进行了很多以板片为主题的项目，这些项目的中心目的是与常规设计进程相反的。对许多项目而言，整个设计还是被看做是推动建造和生产的引物。在这个过程中，结构经常显得复杂而昂贵，有时甚至是不美观或者功能失调的。然而，只要通过设计中一些简单细微的改变，这些问题都是可以避免的。通过设计所构成的形式问题和生产、建设的过程确实存在很少的联系，尽管从建造角度来看，瑞士被看做是这方面建筑的指明灯。然而，随着工具和加工技术的飞速发展，建筑师们要与之保持齐头并进是很困难的。因此为了在建造中设计出新的形式，我们的实验将材料及其加工处理放在第一位。正如在汽车制造业中，过程的规划是生产出简化的、代价合理的产品的一个重要前提。

我们工作室的大部分学生作业显示出，通过这个途径可以形成一些有趣的新型结构，例如2006—2007学年生成的挡风屏障（折叠的屏幕）。

The central goal of all projects that were carried out in our department based around the theme of sheeting is the reversal of the usual design process. For many, the completed design is still seen as the primer from which construction and production is derived. From this process, structures often emerge that are complex and costly, and sometimes even inelegant or dysfunctional. All this can be avoided, however, through simple, small changes in design. But such coupling of construction and production processes with the questions about form that are posed by design happens only rarely, even though Switzerland is seen as a beacon of architecture that is particularly well regarded from the construction point of view. However, since the technology of tools and treatments has been developing so quickly, it is difficult for architects to keep abreast. This is why our experiments put materials and the handling of them at the forefront, in order to devise possible new forms of construction. In the automobile industry, for example, this process planning is one of the central pre-requisites for a pared-down, cost-optimized production.

Much of the student work undertaken in our department shows that, when using this approach, interesting new structures can emerge, for example, the paravents (folding

'Alex Reloaded'

'Alphabet'

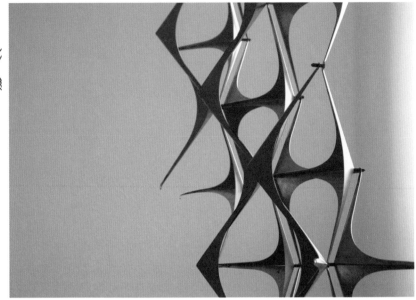

'P.U.R.G.E.'

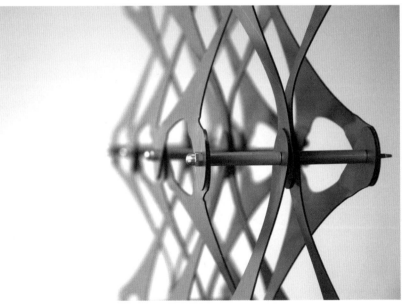

'Node'

这项工作仅仅利用钢板和一个用电脑控制的机器来生产建造元件。其众多成功的形形色色的成果表明，即使受限于单一材料和单一的机器，新的可能性也是很广泛的。

对我们的学生来说，挡风屏障是一次很好的学习体验。尽管结构相对简单，但是单靠材料和机器是做不到这一点的。相对于泡沫板或塑料可以快速简单地被切成块用于制作建筑模型，这些大尺度的结构需要好的实践和技巧来处理。材料和机器的特性不会迎合人的主观需求。在电脑中设计出来的一种很完美的结构，在真正生产时会比预想的花费更多的时间，因为没有考虑到诸多方面的因素，比如材料的预应力、温度、制造误差、工具磨损以及制造顺序等等。如果这种误差不能在整个结构中适当分布，那么这个系统的各个要素就很难组合在一起。我们不能从一个电脑模型中认识到这些问题的重要性，当然就不可能经历这些问题。只有在一个真实尺度下，通过实际机器来对这些实际材料进行操作，才能获得这方面的知识。当然，ETH 没有这些机器，这个挡风屏障是在一个真实工厂中制造出来的，那里的工人都具有相关的专业知识。这样，我们的学生通过这个挡风屏障项目获得的经验，认识到建筑施工的关键所在。

screens) that were generated in the academic year 2006 / 2007. The task was to produce constructional elements using only sheet steel and a computer-controlled machine tool. The many and varied successful results show the broad spectrum of new possibilities, even when one is limited to one material and one machine.

The paravents were a good learning experience for our students. Though the structures are relatively simple, the materials and the machines are anything but that. In contrast to building models using foam board or plastic—which can be relatively quickly and easily cut into parts—these large-scale structures demand a good deal of practice and skill. It is by no means a given that the material and machine will behave the way you want. A structure that exists as a perfectly drawn model in a computer will take much longer to produce than at first thought if the stresses and strains in the material, the temperature, manufacturing tolerances, wear and tear on the tools, and sequence of manufacture are not taken into account. The elements of the system will simply not fit together if the tolerances cannot be distributed properly over the entire structure. The significance of these questions cannot be learned from a computer model; neither can they be experienced. It is only by working with the actual material on the actual machines at true scale that we can gain this knowledge. Of course, we do not have the necessary machines at the ETH. The paravents were produced in real factories by real workers—with commensurately professional results. In this way, our students learned from their experience with the paravents what it means to get to the crux of construction in architecture.

项　目：Freie Innendruck Umformung

时　间：始于 2004

参与者： Oskar Zieta, Philipp Dohmen

合作者： Trumpf Maschinen AG (Baar, CH); Fronius Schweiz AG (Rümlang, CH); Kuka Roboter Schweiz AG (Dietikon, CH); Soutec Soudronic AG (Neftenbach, CH)

吹片 Blowing Sheets

这项科技的基本原理是由在汽车工厂里被称做内部压力成形或液压成形发展而来的。管子是在高压作用下成形的——材料本身几乎是膨胀成形。例如汽车车身的门支柱，就是这样成形的。这个过程非常快速、经济、精确。然而，其实际的成本在于形式的加工，因此不适合应用于短期的建筑。所以，在工作室里，我们喜欢以"尝试一次"的模式思考。

因此，工作室做了有关压力成形的实验，但是并没有利用机器加工。我们将这个新获得专利的过程称作FIDU，即自由的内部压力成形。这个过程中我们没有利用已经定型成一定形状的管子，而是将钢板按照特定的平面轮廓焊接。因此三维的形式不是机器雕刻出来的，而是与激光焊接的轮廓相关。我们因此也具备了生产这种非常快捷而经济的一次性产品的可能性。然而有很长一段时间，在保证可预知性和三维模型再生产方面遇到了一定的问题。我们发现只有建筑师才能捕捉到刚开始粗糙但完备的结构的内在魅力。这或许也可以解释为什么即使这种方式提供了变化空间很大的成形的可能性，包括技术和美学的

The basic principle of the technology developed is known in the automobile industry as internal pressure forming, or hydroforming. Tubes are pressed into a form under high pressure—the material itself is almost inflated. This is how the door pillars of car bodies are formed, for example. The process is extremely quick, economical, and accurate. The real cost, however, is incurred with the machining of the forms. It is not, therefore, applicable to short-run, architectural applications. All the more so since, in the department, we like to think in terms of one-offs.

For this reason the department has experimented with pressure forming, but without the use of the machined forms. We call this newly patented process FIDU, or freie Innendruck-Umformung (free internal-pressure forming). Our process does not use tubes that are then shaped in a form, but rather sheet steel that is welded along a particular contour. In this way it is not the machined form that is responsible for the three-dimensional result: the laser-welded contour is. Therefore, we afford ourselves the possibility of an extremely fast and economical one-off production. For a long time, however, guaranteeing the predictability and reproduction of exact three-dimensional results posed considerable problems. One really needs to be an architect to see the charm inherent in the first rough-and-ready structures that were produced. This might also explain why engineers have so far

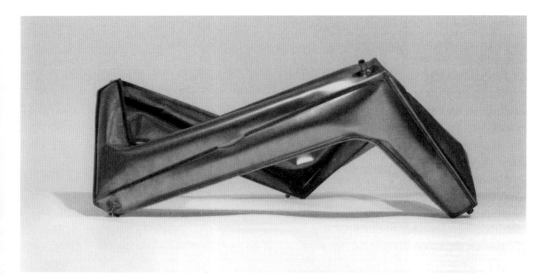

利用自由内部压力成形技术（IPFF），将两片非常薄的钢板切割成需要的形状再焊接在一起，并且利用空气压力使其膨胀。内部空间的扩张使得钢板再次变形，形成了"钢板——天生的张力模型"，所谓的最坚硬复杂的三维结构。

Using internal-pressure free forming (IPFF) technology, two extremely thin sheets of steel were cut to the required shape, welded together, and inflated, using air pressure. Expansion of the interstitial space deforms the sheet steel, forming a complex three-dimensional structure that is then optimally stiffened according to the 'sheet steel-born tension model.'

大尺度的 IPFF 实验，2007。

A large-scale IPFF experiment, 2007

特殊需要，但是迄今为止机械工程师还没有致力于这种类型的建造过程。我们工作室经过许多实验和多轮尝试，研究出了一类"变形字母表"。这表明，各式各样的变形和最终成形之间的关联有时很有趣，有时很意外，这也说明了即使在一个受限的数字链中，创作的自由也是可能的。

具体而言，这个过程进行流程如下：先将两片钢板沿着设计好的轮廓焊接在一起，然后再切割。而这种操作的理想工具就是激光，因为它在切割和拼合方面有相当的灵活性，而且能够直接由电脑控制。其实钢板之所以能够变形，是由于其几何结构和钢板厚度的作用，当然还有外力的影响。而过程中是利用水压还是气压，主要依赖于形状的尺寸。这两种媒介最重要的不同之处在于它们具有不同的密度和可压缩性。使用空气媒介必须保持低压，而且为了确保过程安全，必须输入精确数量的气体。在这方面，水压的安全隐患小一些，但是由于尺寸变大后，水自身重量也会变得非常大，因此水压并不适用于尺寸较大的形状。这种可控制的技术和材料之间的相互影响，使得最终结果得以成形，并且在所有钢板项目中凸显出了显著的特征。

通过第一次实验得出的最重要的结论，与结构的功能性和造型的美观性关系同样密切。由这套流程生产出来的形体证明比由弯片制成的类

not been interested in this type of construction process—although it offers an enormous variety of formal possibilities, including special technical and aesthetic appeal. After numerous experiments and trial runs in the department, a sort of 'alphabet of deformation' emerged. It represented—sometimes playfully, sometimes almost ironically—the interrelationship between the various deformations and the forms that emerge from them, and also illustrates the creative freedom that is possible even in a closed digital chain.

In detail, the process proceeds as follows: Two pieces of sheet steel are welded together along a calculated contour and then cut out. The ideal tool for this is the laser, since it cuts and joins with equal flexibility and can be controlled directly by computer. The way in which the sheet steel actually deforms is a function of geometry, the thickness of the sheet and, naturally, the force of pressure applied. Whether the process uses water pressure or air pressure is highly dependent upon the size of the form. The most important difference between the two 'media' results from their different densities and compressibility. The air introduced must be at a low pressure, and it must be fed in accurately metered amounts in order to keep the process safe. In this regard, water pressure is less dangerous, but is not as well suited for larger forms, since its own weight quickly becomes too great. This interplay between controllable technology and materials, which is finally responsible for the resulting form, acts as a defining characteristic for all sheet steel projects.

The most important conclusions from the first experiments had as much to do with the constructional functionality as with aesthetic appearance. The forms produced by this

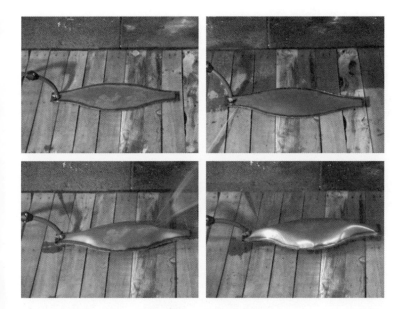

在第一系列由内部压力作用发生自由变形的实验中，钢板仍然由水压作用变形，在此过程中对最终形状稍加控制。

In the first experiments with internal-pressure-driven free forming the sheet steel was still being deformed using water pressure, a process that gave little control over the final shape.

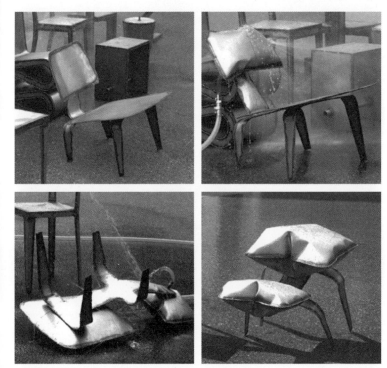

我们的学生对于利用 IPFF 技术对经典设计进行再创造乐此不疲（这里是 Charles Eames 的"木躺椅"）。

Our students had great fun with the experiments to reproduce design classics using the IPFF technology (here the 'Lounge Chair Wood' by Charles Eames), by inflating them.

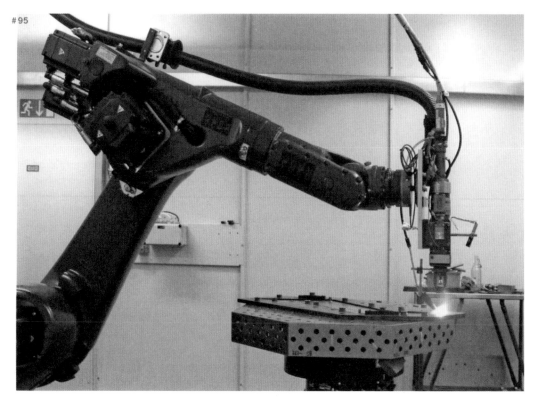

钢板被切开后，由焊接机器人将其焊接到一起，从而使两块钢板之间形成密闭空间。

After sheets of steel are cut, they are welded together using a welding robot, which creates an airtight space between the two sheets.

原型可以很快地由纸片做出来。对比纸模型，钢板在充气后仍能保持它的形体。

Prototypes can be quickly put together from paper. In contrast to the paper models, the sheet steel retains its form after inflation.

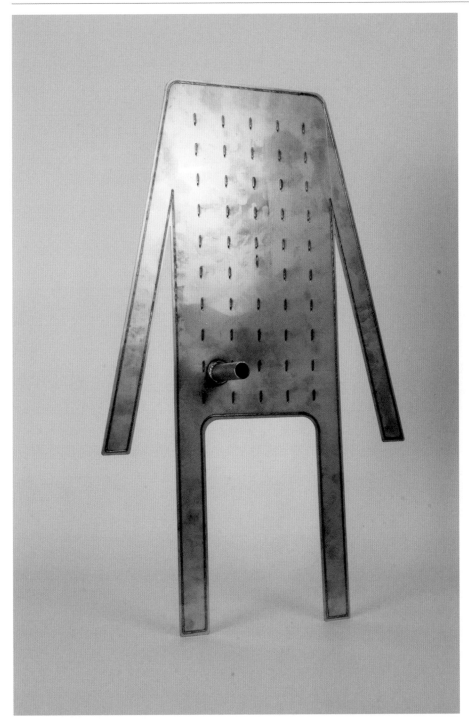

已经焊接好的钢板片,即将被充气。 The welded pieces of sheet steel, waiting to be inflated.

这把名为"Plopp Hay"的金属凳,是利用 IPFF 技术生产的代表作品。它已经赢得了众多奖项,并且售出数千件样品。

A chair – the 'Plopp HAY' bar stool – is the most prominent product to date to be produced using the IPFF technology. It has won many design prizes and several thousand examples have been sold.

似产品更稳定，并且从结构和机械的角度来看，超过了所有预期。例如，6米长的步行桥，如果利用内部压力成形生产，170千克的桥能支撑2 000千克的重量，超过了其自重的十倍。

更多复杂的设计通过略微与众不同的设计语言和外形使观看者眼前一亮，例如Chippen-steel chair（由气压成形的钢板制作而成，已经在Manufactum的目录中备受关注），抑或是在红点设计中取胜的"Plopp"凳。初次看到的时候，这些设计运用的真实材料很难被识别出来。这些材料与形体之间似乎存在着特殊的张力，但是一旦你碰触到其表面或是轻敲它，这种张力也会随之消失。这些设计看起来仿佛是用柔软的材料制作而成，而实际上却是由坚硬的钢板组成的。

我们在实验中得以不断地发现和学习制作加工过程中各个新的方面的细节。以钢板为例，和我们通常所认为的相比，钢板看起来是一种更为多样化的材料，并且钢板弯曲的方向对它们的形变有决定性的影响。

在建筑中，将钢板作为一种既精准无误又灵活经济的承重材料的可行性还没有完全被挖掘出来。例如我们可以想象的，将独立的板材运送到基地，然后再将它们加工形成最终的形体。这种简单的运输和快捷方便的装配将使临时结构受益匪浅。

process proved considerably more stable than comparable objects made from bent sheet, and exceeded all expectations from a structural and mechanical point of view. For example, a six-meter-long pedestrian footbridge, produced using free internal-pressure forming and weighing around 170kg could support a weight of almost 2000kg—more than ten times its own weight.

More complex designs—like the Chippen-steel chair, made of air-formed sheet steel, which has been featured in the catalog of Manufactum, or the 'Plopp' bar stool, winner of a red dot design award—surprise onlookers with their somewhat unusual design language and appearance. At first glance, the true materiality of these objects can hardly be identified. There seems to be a peculiar tension between material and shape that disappears as soon as one touches the surface, or taps it. The object seems to be made from soft material, but reveals itself to be made of hard sheet steel.

During our experiments, we were able to continually discover and learn new aspects of the details of the manufacturing process. For example, sheet steel appears to be a less homogeneous material than is generally accepted, and the direction of bending of the sheets has a decisive influence on their deformation behavior.

The possibilities of employing sheet steel in architecture as a load-bearing material that is precise, flexible and economical to work with have not yet been exhausted. We can imagine, for instance, transporting individual sheets to site, and only then working them into their final form. The hassle-free transport and the quick, uncomplicated assembly would be qualities from which temporary structures, for example, could profit.

项 目: Monster Structures
时 间: 2007
参与者: Kerim Seiler
合作者: Tom Pawlofsky, Benjamin Dillenburger, David Sekanina, Kai Rüdenauer, Steffen Lemmerzahl

由纸板制成的屋顶 A Roof Made of Cardboard

在日常建筑学院的教学过程中，瓦楞纸板都发挥了很大的作用。与别的材料相比，它的成本几乎为零，用它来操作能够非常快速，并且它是一种建筑性很强的材料，甚至可以直接用它来建造物体，即使制作1:1比例的模型，它依旧能够起作用。同时从建设的角度来看，瓦楞纸板的使用比钢板容易得多。然而，它自身的建造方法需要有一套不同于钢板或木材的数字技术体系。总之，它为学生的工作提供了一种有趣的前景。

这个项目的出发点是一个由瑞士艺术家Kerim Seiler的作品。2006年时，他是我们高级硕士研究课程的学生，设计了一个复杂、相互交织的屋顶结构。在"怪物结构"系列研讨会的工作指导下，这个项目的目标是在一周时间内完成包括建设、生产、组装在内的所有工序。由于Seiler一开始就在数字化的基础上工作，我们可以确定有精确的数据来控制所有构件的数量、角度、高度以及直径等等，不需要用手绘去表达任何一个部分。

一项特别的挑战是如何折叠瓦楞纸板，使其结果一方面看起来像是一个木构建筑，另一方面需要通过尽

Everyday corrugated cardboard has many advantages in the education of architectural students. In comparison to other materials, it costs almost nothing, can be worked with very quickly and in large scale, and it is a very architectural material—even though anything built with it, even at a scale of 1 : 1, is still very much a model. And from the point of view of construction, corrugated cardboard is far easier to work with than, say, sheet steel. It demands, however, its own grammar of construction, a differently configured digital chain than that of wood or steel sheet. All in all, it presents an interesting prospect for student work.

The starting point of this project was a piece of work by the Swiss artist Kerim Seiler. In 2006, when he was a student in our Master of Advanced Studies program, he designed a complex, intertwined roof structure. Working within the 'Monster Structures' seminar series, the goal of the project was to build a roof structure within a week, incorporating construction, production, and assembly. Since Seiler worked digitally from the outset, we could be sure of having exact data on the number, angles, heights, and diameters of all the elements, and no part would have to be drawn by hand.

A particular challenge was folding the corrugated cardboard so that, on the one hand, the results would look like a wooden construction, and,

可能少的折叠来组合这些构件。这样做是为了可以承受节点处较高程度的弯曲。整个雕塑的所有构件,我们只用了两种纸板模具,用数控切割机花了两天时间生产出来。这种绘图、切割于一体的控制软件已经被纳入参数化的 CAD 程序,这种打印机驱动程序可以自动将数据转化成机器可读的代码。雕塑的实际装配任务在一天内就完成了,这要归功于各个部分自动生成的识别标记和精确的生产计划,这是一个快速、精确且经济的过程。这个复杂的屋顶建造所花的生产和材料费用,由专业设计师估算的 10 000 欧元,减少到 450 欧元(除去学生参与的工作量)。通过这个项目可以证明瓦楞纸对于学生制作 1:1 规模的建筑结构试验来说是一种很理想的材料,它快速、经济,容易获得,方便处理。我们向其他学校强烈推荐这种材料。

2007 年,我们再次与 Kerim Seiler 合作,进行了一个类似规模的项目:MgBeth,它不是用瓦楞纸板制作,而是使用布和空气。对于这个充气雕塑来说,数字化程序的各个阶段都必须重新设计和重新定义。在空气压力的作用下形状会发生怎样的变化?我们如何利用空气层保持其稳定性?切割与拼接模式是由计算机确定的,切割机用来加工所有的构件。这个雕塑在苏黎世 Migros 公司高层建筑上正式亮相,是由 450 平方米面积的材料和 1.5 千米长的纱线制作而成。

on the other hand, the pieces would be brought together using as few folds as possible. This was done in order to accommodate the relatively high bending moments at the joints. The production of the entire sculpture, for which we used two pallets of cardboard, took only two days, using a computer-controlled cutting machine. The control software for the plotter /cutter was integrated into the parametric CAD program: a sort of printer driver that automatically turned the data into machine-readable code. Actual as-sembly of the sculpture was completed within a day, thanks to the automatically generated identification marks on the individual parts and a precise production schedule: a quick, accurate and, above all, economical process. The production and material costs for the complex roof construction—estimated by professional stage designers to be of the order of €10,000—came to exactly €450 (factoring out the working hours of the students taking part). With this project, corrugated cardboard proved itself to be an ideal material for student experiments in the field of constructive architecture on a 1:1 scale. It is quick, economical, easily available, and easily disposed of. We can highly recommend it to other schools.

In 2007, again with Kerim Seiler, we carried out a project on a similar scale: MgBeth was not built out of cardboard, but was implemented using cloth and air. The individual stages of the digital chain had to be redesigned and redefined for the inflatable sculpture. How would the shape change under the effect of air pressure? How could we use air chambers to influence its stability? Cutting and stitching patterns were prepared by the computer, and cutting machines produced the necessary pre-cut pieces. Four hundred and fifty square meters of material and 1.5km of yarn were used for the sculpture, which was unveiled on the facade of the Migros high-rise building in Zurich.

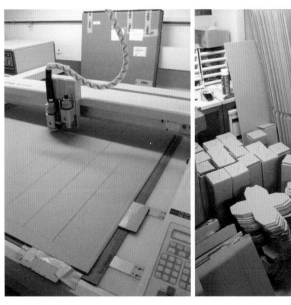

产品数据被发送到电脑控制的切割机上进行自动切割。

The production data was sent to the computer-controlled cutting plotter for automatic cutting.

卡板，作为一种既廉价又轻便的材料，可以用简单的连接技术扣在一起。

As a material, cardboard is not only cheap and extremely light, it can be fastened together using simple jointing technology.

"Tetrascope"是一个由卡板制成的雕塑,由瑞士艺术家 Kerim Seiler 设计,并与我们的学生共同完成,生产装配仅仅用了一个星期。

'Tetrascope' is a sculpture made from cardboard by the Swiss artist Kerim Seiler that, together with our students, we produced and assembled in only a week.

定时记录我们的装配过程。所有纸板的屋顶结构由几个学生仅仅花费几个小时就组装好了。

A time-lapse view of the assembly process. The entire cardboard roof structure was assembled by the students within a few hours.

"MgBeth" 是由瑞士艺术家 Kerim Seiler 设计的另一个雕塑,运用了与 "Tetrascope" 相同的数据结构。然而这次结构是由布呈现出来,而非纸板。学生们在一个飞机库里切了数千米的材料,并且缝制成复杂的结构。

'MgBeth,' another sculpture from the Swiss artist Kerim Seiler, uses the same data structures as 'Tetrascope.' This time, however, the data was 'rendered' not in cardboard, but in cloth. In a hangar, the students cut many thousand meters of material and sewed them together in a complex pattern.

"MgBeth" 在生产车间里进行充气实验。

'MgBeth' during a test inflation in the production hangar.

这个结构被置于苏黎世 Migros 公司的高层建筑上展示，非常显眼。

The sculpture was displayed prominently on the Migros high-rise building in Zurich.

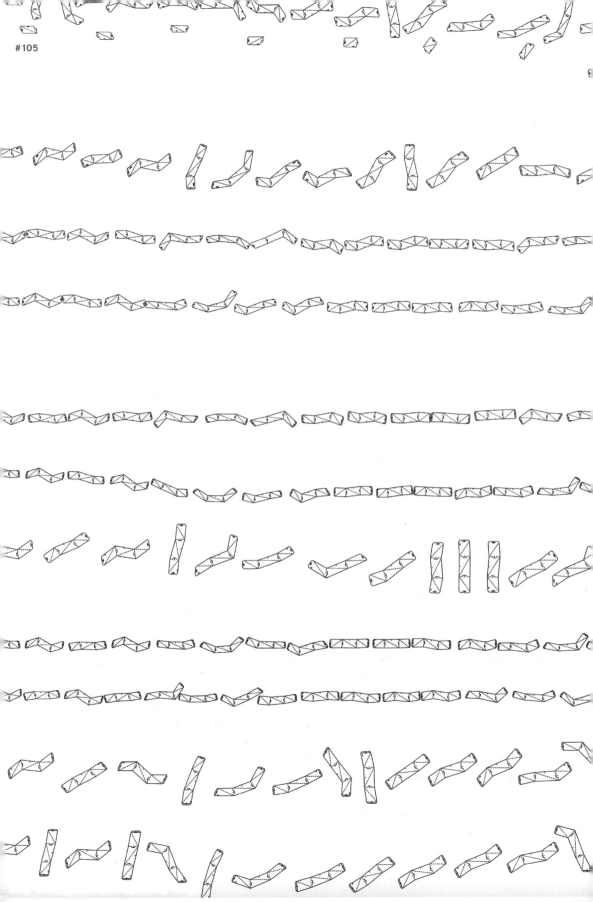

第六章

数据和信息

　　数字革命已然改变了我们主要的世界观：它不再是明确的因果关系，也不是外部的分配或者精确的参考。现在，它主要是关于循环、平衡、交换、竞争和发展。这就是为什么我们需要去理解信息技术何以对建筑如此重要。我们怎样才能在设计中寻求新的竞争和发展的平衡？这样的网络和交流过程应该遵循什么样的规则？

　　首先，我们需要将数据和信息区别开来。没有将信息传递给接受者，换言之就是没有做出任何修改，这种单纯的数据传输就是无效的。如果一个人对汉语（原版为"德语"）或者英语一窍不通，那么即使他读这本书的时候都会得到许多数据，但是至少在本书备受关注之前，他却不能从中学到任何东西。即使是对那些掌握了汉语和英语的读者而言，能够从中吸收多少信息也取决于他们对我们所做的工作的了解程度。更通俗地说：数据转换的信息依赖于接受者，而且必然与实际的数据源有明显的差异。正如大家所知，全世界的可用数据与日俱增——到处都充斥着数据。我们注意到，例如手机、因特网等各种新的交流与行为形式由一个明确界定的 meta-plane 构成。在这里，我们都备受挑战与竞争。也正是在这里，新的缺点和新的价值显露出来。

　　对那些基于信息技术的建筑而言，这种差异至关重要。正如本书中经常提到的，如果越来越多的主持人角色由未来建筑师来承担，那么他们将必须理解交流的规则——这与传统的系统是有本质区别的。计算机辅助的数字建筑试图设计出交流的过程，由行为模式来确定并调和。像维基百科这样的网站项目也是通过同样的方式发挥作用的。创办人和管理员可以建立这样的系统，并且明确规章制度，进行监测，但是却没有人能够让系统完全独立地运行。换句话说，我们可以将这个过程中建筑师的角色比做灌溉、修剪树木的盆栽园丁，却不可以断言本质上谁是整个成长过程中的观测者。建筑师无需非常深入地研究编程，了解编程的原则和信息技术的观念模式足矣。

　　可以通过两个方面衡量计算机辅助系统的质量，即运作过程的规则构想和最终目标。这又将我们带回到了本章节开篇所提及的数据与信息之间的差异。"艺术品"由设计数据模型和个体特征组成，信息也由此产生，从而整个系统通过尽可能少的、优雅制定的、具有目标导向的规则启动。以 Monte Rosa Chalet 的建造为例，这个"艺术品"包含了个体化的建筑元素——这里主要是指床和墙，这些个体元素可以独立地相互作用，从而实现最终目标：减轻建造元素的重量。这种个体元素的机动性和概念上的机动性是一样的：它们以何种方式、向哪个方向，才可以沿着虚拟的路径传输，最终到达目标。而这个目标就是生成一种比其他任何常规设计工具所能得到的更好的解决方案。

　　建筑与信息技术的调节自然是一个挑战。控制项目的规则和目标而非真实世界中的结果，对建筑师来说很困难，因为他们要重塑自己的创造力。毫无疑问，不是以传统的感觉来设计建筑，而是对建筑进行编程形成一套可行方案，将使建筑产生深远的变化。然而，与一个信息技术专家交流建筑问题非常困难，甚至是不可能做到的，当对方是以一种比较轻松、务实的方式进行计算时就更是如此（而在我们的文章中，这恰恰是唯一一种可能实现的方式）。从理论上说，某些算法或许不能提供完全规则的产品或者非常精确的解决方案，但就我们的需要而言，它们也是非常有用的工具。我们一直在寻找的，并不是唯一的完美的解决方案，而是那些能够将电脑运算降低到易管理水平的、有用的、可实践的、高效率的解决方案。

Data and Information

The digital revolution has clearly shifted the emphasis of our worldview. It is no longer about definite, causal relationships or about external allocations or precise references. Now, it is about cycles, balance, exchange, competition, and development. That is why we need to understand how information technology can be so essential to architecture. How can we design in balance and initiate competition and development? What rules do networks and communicative processes follow?

Firstly, we need to differentiate between data and information. The pure transmission of data is ineffective if it does not deliver information to the recipient—in other words, if it does not change anything. If one has no understanding of German or English, anyone reading this book would certainly understand a lot of data, but—at least as far as the text is concerned—would derive no benefit from it. And for those readers who can speak German or English, the information depends upon how much they already know about our work. More generally put: the information of data exchange is dependent upon the recipient, and must be clearly differentiated from the physical data flow. As everybody knows, the amount of data available worldwide is growing rapidly—there is a glut of data. The new forms of communication and behavior that we can observe, for example, with mobile telephones and the Internet, are emerging from a clearly defined meta-plane. Here, we are all challenged, we are all in competition. It is here that new shortcomings and new meanings emerge.

For architecture that is based on information technology, this difference is of central importance. If, as we have often claimed in this book, the role of moderator is taken on more and more by future architects, they will have to understand the rules of communication—and these are fundamentally different from the classical system. Computer-aided, digital architecture seeks to design communication processes, to set them in motion, and to moderate them. Web projects like Wikipedia function in exactly the same way. There are founders and administrators who set up the system, define the rules, and carry out the monitoring, but who then let the system run independently. Put another way, one could say that the role of the architect in this process could be compared with that of a bonsai gardener: someone who waters their trees and prunes them here and there, but who is essentially just an observer of the growth process. Architects won't even have to plunge into the depths of programming. It will be enough to understand the principles of programming and the mindset of information technology.

The quality of computer-aided systems is defined by two aspects: formulation of the rules of play and the goals. This brings us back to the difference between data and information that we mentioned at the beginning of this chapter. The 'art' will consist of designing the data models and the individual parameters so that information actually flows, so that the entire system starts up with the smallest possible number of elegantly formulated, goal-directed rules. For example, for the construction of the Monte Rosa Chalet (The Digital Chalet → pp.138 ff.), the 'art' consisted of allowing the individual building elements—

here they consisted mainly of beds and walls—to interact with each other in such a way that the individual elements could behave with sufficient independence to achieve the overall goal: reducing the weight of the construction elements. Defining the mobility of individual elements in such a way is equivalent to a conceptual mobility: how, and in which direction, they would be transmitted along the virtual way to reach their target. And the target is to generate a solution better than any that could have been achieved with conventional planning tools.

    The mediation between architecture and information technology is, naturally, a challenge. The new dynamics which allows to control the rules and the goals of the project—but not its real-world results—is as difficult for architects to grasp as the fact that their creative abilities are being reallocated and altered. To program buildings as an ensemble of possible solutions instead of designing them in the classical sense will be, without doubt, a far-reaching paradigm change for architecture. Conversely, it is also difficult—maybe even impossible— to communicate architectural problems to an information technology specialist, even more so when one has a comparatively relaxed pragmatic approach to algorithms (which, in our context, is the only way to proceed). Even if certain algorithms—from a theoretical point of view—may be unfit for the production of a mathematically clean and analytically precise solution, they can, for our needs, be extremely useful tools. This is because we are not always looking for the one-and-only 'perfect' solution, but rather for useful, practicable, and efficient results that keep the computer time down to manageable levels.

Jacob Bernoulli
Ars Conjectandi

' If besides the arguments that count in favor of the thing,
other pure arguments present themselves,
which indicate the opposite of the thing,
the arguments of both kinds must be weighed separately
according to the preceding rules,
in order that one may obtain a ratio between the probability of the thing
and the probability of the opposite of the thing. '

立面

Facades

项　目：Südpark（瑞士，巴塞尔）

时　间：2005—2008

参与者：Steffen Lemmerzahl, Ludger Hovestadt

合作者：Herzog & de Meuron (Basel, CH)

从内部到外部 From Inside and Outside

立面设计事实上成为我们 CAAD 研究项目的主要应用领域。我们将用很多例子来说明数字化规划和建造过程在建筑实践中的应用：通过打破仍然很强地影响着当今立面设计的固有格网的约束，或者在设计过程中重新获得自由的空间排布，虽然这在不久之前被认为是不可能的。用我们的方法可以重新定义建筑外观和内部生活的相互作用。立面在两方面起着作用：对外，立面设计中的美学、设计个性以及对环境的融合都发挥着它们的作用；对内，立面结合了提供景观视线、隔热和隔绝噪音等不同功能。

与赫尔佐格和德梅隆建筑事务所合作完成的一个项目中，我们证明了立面的设计及实施可以发挥怎样的作用。这个项目是巴塞尔总火车站附近的一个大型建筑综合体，立面设计由六种模数控制，六种模数经过设计其比例可以自由变化。一方面，这些立面模数满足各种各样隔绝噪音和采光的功能需求，并且在城市文脉的框架内为居住者提供了景观视线。另一方面，这些立面模数解决了由于不同的功能使用——零售、办公空间、公寓等——所带

Facades have turned out to be a major field of application for our CAAD research program. We have used many examples to illustrate the application of digital planning and construction processes in architectural practice: by breaking out of the confines of the rigid grids that still strongly influence present-day facade design, or by regaining free space in designs, thought impossible only a short time ago. The interplay between the exterior appearance and the interior life of a building can be newly defined using our methods. Facades work in two different directions: externally, where aesthetics, design individuality and integration into the environment all play a role; and internally, where they incorporate functions such as providing views, and thermal and noise insulation.

We demonstrated how design and implementation of a facade can function in a project carried out in conjunction with the architectural practice of Herzog & de Meuron. The project consisted of a large building complex in the area of the main station in Basel, where the facade was adorned with six differently designed and scalable modules. On the one hand, these facade modules dealt with various noise insulation and illumination functions and also provided views for the inhabitants, all within an urban context framework. On the other hand, they had to cope with the various challenges posed by the diversity

Südpark (Basel, CH)

瑞士的赫尔佐格和德梅隆建筑事务所设计的立面。变量规则系统生成了一个结构上类似但形式上却不规则的立面。建筑师不是直接地提出了形式的结果,相反,这是一个规则系统的产物。

A facade by the Swiss architecture practice of Herzog & de Meuron. The parametric rule system generated a facade that is structurally homogeneous but formally irregular. The architects did not produce the formal result directly; rather, it was a product of the rule system.

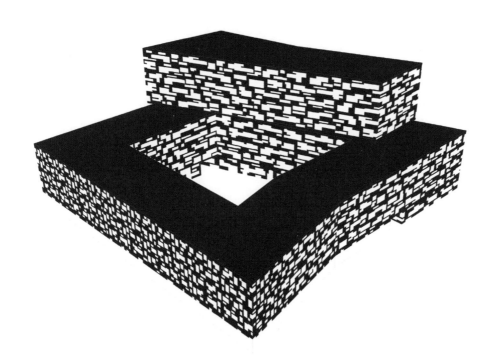

通过 162 次尝试,确定了最终规则和变量,获得的结果实现了建筑概念。

After 162 attempts, final rules and parameters were found, leading to a formal result that fulfilled the architectural concept.

来的各种挑战。建筑师要面对的问题很明显：该设计不能简单地用手工设计来控制。

为了用程序生成立面，我们不仅需要这些模数的几何数据，还需要了解几何形体上的可能变化、城市设计中要用到的参数，以及在建筑上的应用。以这些数据为基础，我们和建筑师一起形成了一套系统的规则，用来选择和调整特定的立面元素。电脑决定剩下的内容：模数化立面序列的运算、它们相应的比例和位置、各种立面的视觉效果，并且同时自动绘制出轴测图供使用者查看。该过程十分有效，我们得到了162种不同的立面，该数字还在不断增长。当我们微调了程序框架后，最终得到建筑师满意的结果。

在这个过程的最后阶段，我们不仅有一套完整的建筑各元素的数据——包括立面上开窗的数量，以及所有窗洞需要的线性数据，我们还有所有建筑元素的清单，以及它们自动生成的实现方案，我们可以把这些数据直接输入到一个由电脑控制的施工程序中。

这些立面设计项目为我们提供了非常有价值的想法，并为数字与传统方法的结合提供了一条简单的途径。

of use: retail and office space, apartments, etc. The problem for the architect was clear: the design could not be manually controlled.

In order to program the facade, we needed little more than geometrical data for the modules and their geometric possibilities for adaptation, the parameters of the urban design contexts, and the applications for the building. Using this data as a basis, we formulated, with the architects, a system of rules whereby a particular facade element would be chosen and adapted. The computer determined the rest: the calculation of the sequence of modules, their scaling and their corresponding positioning, and the visual appearance of the various facades, as well as automatically generating isometric and perspective projections. The process was so efficient that we were able to permit ourselves the luxury of developing 162 versions of the facade, until, after fine-tuning of the framework conditions, and formulating the system of rules and the design of the modules, the architects were satisfied with the end result.

At the end of this process, we not only had a complete set of figures for all the building elements—including the numbers of openings in the facade or the distance, in linear meters, needed for all the window frames—but we also had a complete schedule of every building element employed, along with automatically generated implementation plans for every single element, which we could feed directly into a computer-controlled production process.

These facade projects provided us with valuable insight and allowed easy integration of our methods into the established design processes of traditional

通过数字链方法，我们可以增加设计过程的价值。

现在，立面设计可以被实时地反应，并且能很实用地应用到最后的细部中。这是因为——从规划、建造和设计的角度——电脑，就像一个 CNC 机器，生产单个个体和量产没有什么区别。建筑中应用信息技术的真正挑战并不在于技术本身。就我们参与这些或其他类似项目所积累的丰富经验而言，挑战更在于如何将我们的思路和方法告诉甲方。幸亏有数字技术，一个高度个性化却又功能明确的建筑可以在我们的控制中设计出来，并进行投资施工。

architectural practices. And by using the digital chain we could add real value to the process.

Today, the design of facades can be simultaneously adapted and functionally tuned to the last detail. This is because—from the point of view of planning, construction, and design—a computer, like a CNC machine, makes no distinction between individual and mass production. The real challenge for applied information technology in architecture is not the technology itself. On the basis of our wealth of experience with these and similar types of projects, the challenge lies rather in communicating to all project stakeholders that, thanks to digital technology, a highly individualized and yet functional architecture can be controlled, implemented, and financed.

项 目：	**Alu Scout**
时 间：	2006
参与者：	Steffen Lemmerzahl, Alexander Lehnerer
合作者：	Alu-M. AG (Münchenstein，CH)
赞助者：	Eduard Hueck GmbH & Co (Lüdenscheid, D); Verband für Oberfächenveredelung von Aluminium (VOA) (Nürnberg, D); Alu-M. AG (Münchenstein, CH)

参数化的立面 The Parametrical Facade

当我们用正确的软件进行电脑编程时，设计出新立面的可能性就会大大增加，正如我们在"Alu Scout"中所做的那样。在 CAAD 研究项目中的同学参与了"勇敢的裁缝"工作组，他们为一个铝业工厂举办的年度创新竞赛做了好几个设计。2006 年，这个竞赛的主题是铝材和玻璃立面。

所有项目统一的主题就是参数化的立面。在这个课题中，"参数化"意味着一个立面中单个的元素从构造和结构的角度来说是一样的，但是根据环境或使用者的要求，它们的最终形式并不相同。当然，它们最终的形式不是由人来直接决定，而是由一个有着特定参数的程序来决定。这种工作方法最令人吃惊的地方在于：尽管设计过程中有着相对严格的材料限制（比如说挤压铝），以及一个非常严格的设计要求，我们还是能够提出大量多种多样的解决方案。

这里允许我们简短地展示一些项目。有一个工作小组设计一层高的立面构件，每个构件都由6个水平的铝板构成。

The potential for the design of new facades becomes evident when one programs the computer with the correct software, as can be seen in our contributions to 'Alu Scout.' Students in the research program for CAAD took part in the 'Brave Tailor' workshop. They produced several pieces of work for a competition with a prize for innovation held every year by the aluminium industry. In 2006, the subject of the competition was aluminium and glass facades.

The unifying theme for all projects was the parametric facade. In this context, 'parametric' means that the individual elements of a facade are identical from the point of view of construction and structure, but differentiate themselves in their final form according to the demands of the environment or the user. Of course, these final forms are not created manually, but by a program with defined parameters. The most surprising aspect of the work was that, despite a comparatively rigid material like extruded aluminium and a rather narrow brief, we were able to propose a wide variety of solutions.

Space permits us to briefly present a few projects. A working group prepared story-height facade elements, each composed

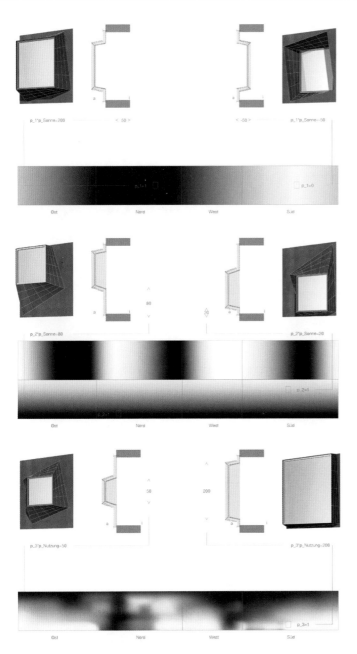

参数化的立面模块在这三个方面可以灵活变化：窗户等级（上图）、窗户在立面上的位置（中图）和窗户的尺寸（下图）。通过适当调整立面上颜色的深浅，例如用 Photoshop 软件，立面上的参数可以被简单和直观地控制。在下图中，浅色表面生成一个大窗，深色表面则是小窗。例如在北面，为了生成较大的窗户，立面只会被标记成浅色的。

A parametric facade module with three degrees of freedom: window levels (above), the position of the window on the surface of the facade (middle), and the size of the window (below). By slightly changing the color density in the wound up facade, for example with Photoshop, the facade's parameters can be easily and intuitively controlled. In the scheme below, a light surface produces a large window, and a dark surface, a small window. On the north side, for example, the facade development would be marked only lightly, in order to generate larger windows.

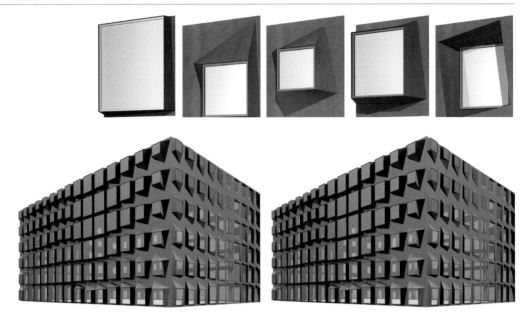

学生完成的参数化立面。参数生成的模块类型和整个立面系统。

Parametric facades created by students. The spectrum of parametrically generated modules and the isometry of the facade.

在施工工地上的 1∶1 的立面构件。

A mock-up of the facade elements at scale 1∶1 on a construction site.

它们的曲率、半径、加工都可以根据光的入射、噪音等级和视线而改变。另一组设计了一个"可伸缩的"模块，它有着可变的窗框厚度，可以根据外部影响和内部需求做出反应。再有一个获奖的项目针对窗户的深度和尺寸编程。第四个项目基于给定的参数，根据板片的宽度、尺寸、是否开启和位置组织了垂直方向规则的板片。还有一个有趣的设计在每个立面构件上有着两个沿同一对角线开启的窗洞，它们的高度、尺寸、位置和深度——包括其他的数据——都可以根据给定的参数进行变化。

应当指出的是这些结果都不是可以手动得到的，因为即便费了很多努力去手动设计，得到的结果并不完善。它们只能通过电脑理性、精确地得出来。这些项目从美学和功能的角度都是令人信服的，并且适用于采用普遍技术的建筑，它们也为我们研究项目的工作提供了一个证明——这也是竞赛委员会一直认可的：五份奖项中有四份授予了"勇敢的裁缝"工作组的同学们。

of six horizontal aluminium sheets. The curvature, radius, and machining could all be interchanged according to the incidence of light, noise levels, or sight lines. Another team created a 'telescopic' base module, with variable frame depth that could react to both external influences and internal needs. Another prize-winning project programmed the depth and size of the windows. A fourth project arranged vertically ordered sheets according to their width, size, opening, and position, based on defined parameters. Yet another interesting design had two window openings per facade element, arranged along a mutual diagonal; the height, size, position and depth—among other things—could be altered according to defined parameters.

It should be pointed out that these results are not achievable manually, because, even after enormous effort, they would remain incomplete. They can only be rationally and precisely defined and produced by computer. Since the projects were convincing both from an aesthetic and functional point of view, and were suitable for building purposes using standard technology, they present a vindication of the work carried out in our research project—something the competition jury obviously agreed with: four out of five prizes were awarded to students from the 'Brave Tailor' workshop.

项　目: Credit Suisse（瑞士，苏黎世）

时　间: 2007

参与者: Steffen Lemmerzahl, Ludger Hovestadt

合作者: Stücheli Architekten (Zürich, CH) mmenarbeit / Cooperation

缠绕的 / 装饰的混凝土　Winding/Wrapping Concrete

数字设计最广为人知的应用之一是在这个大型的建筑综合体中，在这个手工设计的项目中，电脑被当做技术、建造和生产的辅助工具来使用。这个项目是苏黎世 Stücheli 建筑事务所设计的瑞士 Credit Suisse 总部的扩建。为了在视觉上削弱巨大的体量感，建筑师在立面上设计了像细微波浪一样的环绕建筑一周的混凝土带。但是，当考虑到预算时，很明显这一概念不能用传统材料来投资建造，因为几乎每个混凝土构件无论从宽度还是厚度上都是不一样的，因此，它们很难用图纸描绘出来，而且也很难生产出来。

对这些混凝土构件以及用这些构件组成的整个立面（包括玻璃构件）的坐标进行编程，这个问题很快得到了解决：整个数据列表不超过 231 行。正如前面提到的，几乎所有的混凝土构件都是独一无二的，总共有 900 个左右，其中 700 个在体量、高度、厚度、长度和角度上都不同。而在我们看来，这个建筑虽然复杂，但是以现代电脑技术似乎不存在建造难题。然而，从这种现有混凝土构件生产商的角度，这项任务——无论过去还是现在——

One of the most popular applications for digital design is enabling the building of a large, complex project—and, in addition, one that was designed manually—with the aid of computer-aided technology, construction and production tools. In this instance, it concerns an extension to the Credit Suisse headquarters in Zurich designed by the Zurich-based Stücheli Architectural practice. In order to visually lighten the large volume, the architects chose to use bands of concrete on the facade, which encircle the building like gentle waves. However, when it came to the calculations, it was clear that this concept could not be implemented or financed using traditional materials, since practically every concrete element was unique, varying in width and depth, and was, therefore, as difficult to manufacture as it was to draw.

Programming the coordinates of the concrete elements—and with them the entire facade including the glass elements—was quickly resolved: The entire dataset comprised no more than 231 lines. As has already been mentioned, almost every concrete element was unique, totalling around 900 parts, 700 with differing masses, heights, depths, lengths, and angles. A complex construction, but one that—manufactured with modern computer-aided production machines—should not be, in our opinion, a problem. However, from the

Credit Suisse (Zürich, CH)

办公楼立面上预置混凝土构件参数化图集。

A list of the parametrically generated production plans for the prefabricated concrete elements of the facade of the office building.

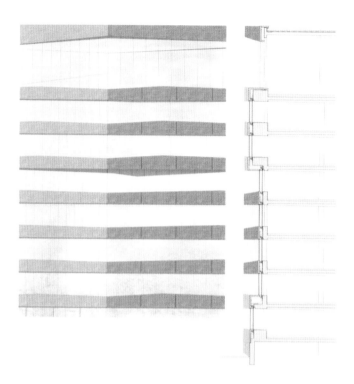

苏黎世 Stücheli 建筑事务所设计的立面和剖面。

The design by the Zurich-based Stücheli architects as a facade elevation and section.

从西面看过去的新建建筑的透视图。

Perspective representation of the new building as seen from the west.

都不是他们能够很容易操作的。在一般的生产过程中，这种构件的模板是由制模机生产的。如果这种模板能够被重复利用，那么从经济的角度来说，这种生产方式或许还有些道理。但是，如果每一个混凝土构件都需要用锤子和锯子来制作一个单独模板，这笔开销就无法控制。这些构件的操控生产已经在本书里充分地阐述过了。不幸的是，我们没有找到合适的合作者，因为在接这种规模的项目时，人们一般都会考虑到风险性。

尽管遇到了困难，但是也许正是这个原因，电脑操控的方法能够在这个项目里证明它的长处和优势。我们只需要相对少的工作去简化和调适整个系统，就可以大量地消减不同构件的种类。新的几何数据被输进电脑，然后重新计算所需的数据——这只需要几个小时的工作，而不是几天甚至几周。将混凝土构件变化的梯度理性化，或对两个单独体块连接节点的简化等方法——这里仅列举其中几个我们采用的几何方法，我们就能将不同构件的数量减少到 66 个。

这个项目也许不那么吸引人，因为最初方法的长处——快速且经济地生产单独的构件——没有被实现。全面地采用数字技术，特别是在大型项目中，仍然需要时间。不过，很明显的是，这些复杂的立面构造在没有计算机技术帮助的情况下将会难以完成。

perspective of the manufacturers of such ready-made concrete elements, this task was not—and is not—something that they could readily manage. In a normal production process, the formwork for such pieces would have to be assembled by shuttering carpenters. This process might make sense from a financial point of view if the formwork could be reutilized. However, the outlay cannot be justified as long as an individual form has to be assembled for each concrete element, using hammers and saws. The computer-controlled production of these pieces has already been fully discussed in this book. Unfortunately, we could find no suitable partner, since, understandably, people tend to be risk-averse when taking on projects of this size.

Despite, or maybe because of this, the computer-controlled approach was able to demonstrate its strengths and advantages in this project. It needed comparatively little work to simplify and adapt the entire system and to reduce the number of differing individual parts drastically. New geometrical data was fed into the computer, and repeated variants were calculated anew—the work of a few hours, not days and weeks. Through measures like the rationalization of the gradients in the concrete parts or the simplification of joints between the individual blocks—to name only a few of the geometrical approaches taken—we were able to reduce the number of different pieces to 66.

The project may be less appealing because the original strength of the method—a quick and economic production of individual pieces—was not realized. The comprehensive adoption of digital technology—especially in complex projects—will still take time. However, it is clear that these complex facade constructions would be unachievable without the help of computer technology.

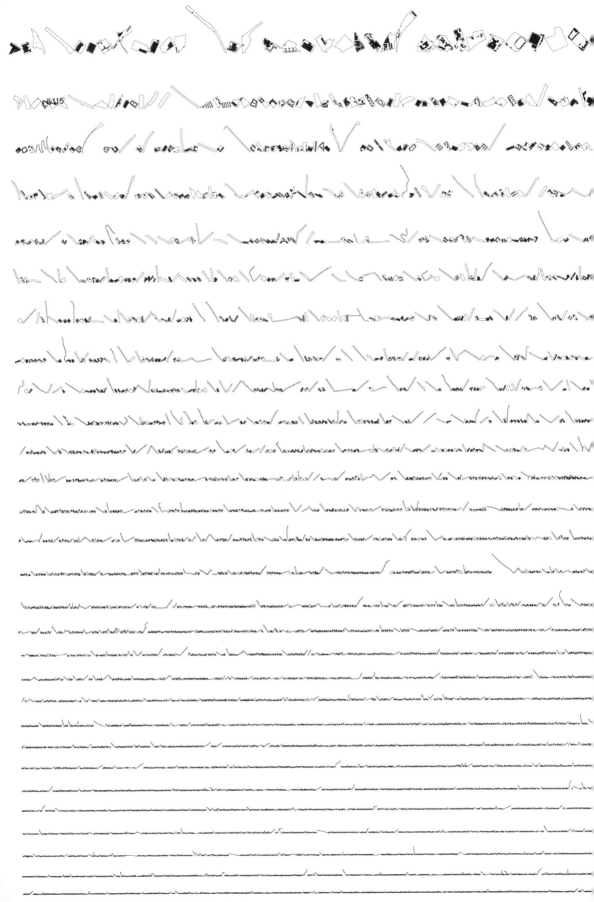

当事物开始学会跑的时候

我们的世界很复杂——非常复杂。很多人尝试通过理论、模型、模仿、格网、定义或机器的方法去控制那种复杂性。当然，也包括用计算机。在第二次世界大战后，伴随着算法正起着越来越重要的作用，信息技术迅速起飞，对于人类掌控事物能力的信任逐渐动摇，因为我们可以把责任交给没有问题的、客观的算法。尽管这种转变过程被严格控制，仍然有人怀疑信息技术，我们是在详细阐述一个主题，一个一开始就排斥这种信息控制的主题。文化和技术常被用来对抗自然。对于这种对抗，有两种相反的态度：一个用现实的责任和控制力作为技术发展的推动力，将"一切皆有可能"作为它的格言。另一个用同样的问题来质疑技术本身，遵循的格言是："回归自然"。

我们在 CAAD 的研究成果通常被理解为一种类似于图解的方式：信息技术的拥护者似乎能将所有事情做得又快又好，然而成立建筑事务所恐怕会失去他们个人创作的自由。总的来说，在这种相互的对峙下，没有任何有意义的结果。

关于技术爱好者的立场，工业基础类（IFC）是个很好的例子。尽管 IFC 仅仅是一个格式，可以无限展开的 XML 格式，然而它们被用于尝试创造一个全世界未来建筑的整体分类方法。更彻底地说，这个著名的倡议认为应该让 IFC 为所有未来建筑提出一种能够描述建筑所有细节的标准化格式。应当重点强调的是未来的建筑，而不是那些迄今为止已经建成的。这一格式会十分灵活，因为它应当像软件一样能被使用。但是，旧标准顽固地存在的倾向广为人知。那么在这试图执行 IFC 标准的 15 年中发生了什么？格式被更加细化，却没什么大的作用。这也为那些不采用这些被精心设计过的格式的规划师提供了批评的根据。因为这样，IFC 仍然没有得到理性化的效果。

这是任何这种类型项目的典型。它们与那些当地的彼此不同的建筑对抗着。而且，因为受到过去强大的文化制约，那些彼此十分不同的建筑还特别排斥改变。这不同于在例如交通领域或机械制造领域的情况，这些学科相对于建筑来说更勇于革新。因此，技术爱好者的立场在建筑中只能引起微乎其微的改变：因为主要基于技术的项目十分有限。

即便新技术与建筑设计形成联系，这一情况并没有太大的改善。让我们想想那些没有电脑辅助的非标准化建筑，它们的几何形式和结构具有实验性，但永远无法实现。在这里，使用电脑不是一个用来挤压调整成本的过程。确实，有创造力的建筑师的角色在这里更加重要，因为通过技术的精炼，他可以获得对几何形更多的掌控。这也是为什么计算机的使用在建筑中不能扮演特别实用的角色。

复杂的几何形体被建筑师市场预算所控制，复杂的结构被规划预算所控制。因此，我们在这个重要应用领域里增加的花费只占不超过建筑总体投资的 1%。这对于在建筑学科内指导性的重要研究和发展来说是不够的。因此，从技术的角度，这些建筑实验相对于其他行业而言——

那些通常比建造业小得多的行业，仍然只是边缘分支！

因此，我们领域的经济形势并不像我们期望的那样。最初在瑞士开展项目时，我们表明不想把建筑放到电脑里，我们想把电脑里的建筑实现出来。我们成功的标准是程序应当证明它们在建筑行业中是有用的，并且它们的实施应当在经济上是可行的。同时我们更加确信，把建筑放进电脑的方法不会发现这些应用。

那么，让我们回到起点，重头再来。

我们的世界很复杂——非常复杂。很多人认为这种复杂性不能被控制。他们认为这个世界不能被完全解释清楚，不能被完全清楚地分割。但是，这一观点并不会无助地投降。他们让事物按照自己的轨迹运转，他们在继续的过程中不断获取成果。当然，这一行为在我们这个时代非常奇怪，但并不是什么新鲜事。我们只需要回顾一下 Leibniz 的单子理论。早在 Norbert Wiener 之前，在没有论述通过控制得到一个更好的世界的控制论时，Leibniz 就让人和机器按照时间的节奏跳舞了。

如果我们回顾巴洛克建筑，或者 Paul Klee 的绘画，或者其他很多在电脑出现之前的例子，我们经常能看到，例如材料、形式或体量都被赋予了它们自己的生命。为了进一步的发展，允许一种偶然的、受保护的内部存在性在具体的情形下表达自我。运用符号系统，我们很早就能精确描述内在如何能表达自我需求，而不需要定义任何内在或外在的大量细节。并不是它们本身需要被精确描述；我们能描述它们的运动或行为就足够了。如同符号机器，这正是电脑的初衷和主要长处所在。在这个关系中我们将技术作为技术爱好者和技术恐惧者的衍生物，可以阐明第三个立场。在这个关于技术、文化和自然的可调和的关系中，人们可以获得显著的经济潜力——这些潜力可以在整合信息技术和建筑的发展中看到。

As Things Have Learned to Walk

Our world is complex—very complex. Many people try to gain control over that complexity by means of theories, models, simulations, grids, definitions, or machines. And, of course, with computers. Information technology took off shortly after the Second World War, as trust in humankind's ability to keep things under control was shaken to its core, and as there was a great sympathy for passing on responsibility to supposedly unproblematic and objective mathematical algorithms. But in spite of the progress in increasing control, suspicion remains that, by doing so, we are simply trying to exhaust a topic, which resists this kind of control in the first place. Culture and technology are happily pressed into service to oppose nature. In this confrontation, there are two extreme stances: One uses the problem of responsibility and control as a springboard to ever more refine technology, taking as its motto 'Everything is Possible.' The other uses this same problem to call the technologies themselves into question, following the motto: 'Back to Nature.'

The role of our research in CAAD is often seen in a similarly schematic way: The protagonists of information technology seem to be able to do everything faster and better, while the established architects are scared of losing their freedom of individual creativity. Generally, in such a mutual stand-off, nothing meaningful results.

For the technophile position, the Industry Foundation Classes (IFC) are a good example. Although the IFCs are simply a format, the infinitely extensible format of XML, they are used in the attempt to create a unitary classification of all of the world's future buildings. More drastically put: This prominent initiative is driven by the idea that for every future building, it should be possible for the IFC to develop a standardized form that can describe the building in all its detail. It is important to emphasize, buildings of the future, not those that have been built to date. The form would be very flexible, because it would be implemented as software. However, the actual tendency of factual standards to persist rigidly is well known. And what has happened in the 15 years spent trying to enforce this standard? The forms have become ever more detailed, something which helps little. It also gives grounds for criticism of planners who do not use the forms, despite the great care with which they have been designed. For this reason, the IFCs have not thus far had their desired effect of rationalization.

This is typical of any project of this type. They are fighting against the non-computability of locally and individually differentiated architecture. Furthermore, architecture—as much as it may differentiate—is particularly resistant to changes, thanks to strong cultural anchors that reach back into the past. This is different for fields such as transport or mechanical engineering, disciplines that are often held up to architects as being friendlier to innovation. Therefore, the technophile position should cause few changes in architecture: Simply because there is little request for projects based primarily on technology.

Even where new technologies form a liaison with architectural design, the situation does not improve much. Let's consider all the experimental geometries and structures of non-standard architectures springing up everywhere and which, without a computer, would never see the light of day. Also here, the computer is not used for a restructuring of cost-intensive processes. Indeed, the role of the creative architect is even stabilized here, through technological refinement, as he gains more control over geometry. This is why the use of computers can play no specifically economic role in architecture.

Complex geometries are financed by the architect's marketing budgets and complex structures by the planning budgets. Therefore, the value added in this important area of application for our department is under 1% of total investment in the buildings. This is insufficient for conducting serious research and development from within the discipline of architecture. So, from a technological point of view, these architectural experiments remain peripheral offshoots of other industries, industries that very often are up to an order of magnitude smaller than the construction industry!

As a consequence, the economic situation of our field is not exactly as one would wish for it to be. At the start of our work in Zurich, we stated that we didn't want to put architecture on computers, we wanted to bring it out of computers. Our criteria for success were applications that would prove themselves to be of use in architectural practice and whose implementation would make economic sense. We have become convinced in the meantime that, using the above-described approaches, we will not find these applications.

So, let's start from the beginning. REWIND.

Our world is complex—very complex. Many people accept that this complexity cannot be controlled. They accept that the world cannot be fully explained, that it cannot be fully parcelled in definitions. However, this stance is not one of helpless surrender. They let things run their course and they 'continue' productively with and among them. Certainly, this behavior is strange in our times, but it is nothing new. We need only look at Leibniz's monads. Long before Norbert Wiener, and without the cybernetic proposition that a better world could be achieved via control, Leibniz had humans as well as machines dancing likewise to the rhythms of time.

If one looks at baroque architecture, or at the drawings of Paul Klee or at many other examples produced long before computers, we see every so often that, for example, materials, forms, colors, or volumes were assigned a life of their own. A contingent, protected interior existence was allowed to express itself in a concrete situation in order to further develop. With symbolic systems, we have long been able to describe precisely how an interiority can express itself needing to determine neither the interior nor the exterior in any great detail. It is not the things themselves that need to be described exactly; it is sufficient if we can describe their movements and behavior. As symbolic machines, this is where computers have their origins and their fundamental strengths. We can formulate a third position in the relationship, with technology as a derivation of technophile and technophobe. In this conciliatory position between technology, culture, and nature, a startling economic potential can be found. Potentials that can be expected from developments with such a fundamental scope as that of integrating information technology and architecture.

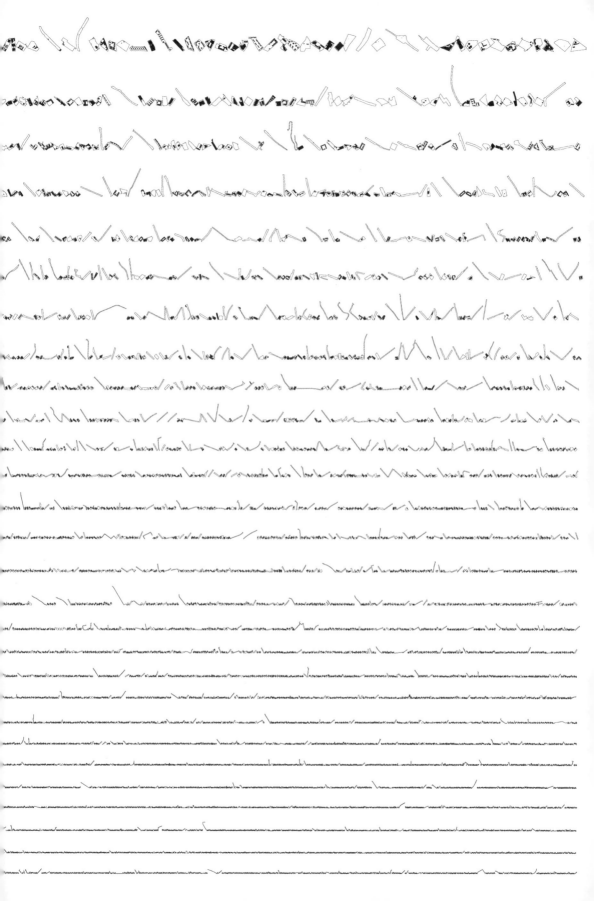

Gilles Deleuze:
Difference and Repetition

' Thus, with actualization, a new type of specific and partitive distinction takes the place of the fluent ideal distinctions. We call the determination of virtual content of an idea differentiation; we call the actualization of that virtuality into species and distinguished parts differenciation. '

全球化设计

Global Design

时　间：始于 2005

参与者：Christoph Wartmann

无处不在的电脑　Computers Everywhere

在 CAAD 的研究项目中，我们正在寻找一种建筑设计的新技术和新方法。最重要的是，信息技术在建筑学的进步中可能扮演了重要角色，这一点促使着我们去寻找。信息技术在建筑项目中有许多用途：从设计、建造、生产到建造自动化中的逻辑计算，以及建造中包含的诸如舒适度、安全性、能耗的调控。虽然我们有许多其他的工具设备可选择，但是数字技术在发展个人技术方面，以及与信息技术人员、电子技术专家和机械工程师们竞争方面有着无可比拟的作用。如果集中于某个特定方面，我们很可能失去广阔的视野。我们关注的焦点依赖于对其他领域的持续关注，并使它们服务于建筑学。我们视自己为建筑师的技术侦察员。

多年后我们发现，和建筑师交流某种技术往往是困难的，即使这种技术在专业的后期发展中有非常重要的地位。有计划地处理 CAD 系统就是典型的例子。CAD 不只是被建筑学学生用来作炫目表现图的工具。然而一些运用于建筑制图运作的概念，诸如图层、类别、块或者语义学模型，绝大部分仍然在他们的兴趣范围之外。

In our research program for CAAD, we are looking for new technologies and methods for architecture. Above all, what drives us is the possible role of information technology in the advancement of architecture. Information technology has many applications within the architecture process: from design, construction, production, and logistics through to building automation, as well as encompassing issues such as comfort, security, and energy. Although we have access to a wide selection of tools and equipment in the department, it would be presumptuous to develop individual technologies ourselves and, in so doing, to compete with information technologists, electronics specialists, or mechanical engineers. By concentrating on very specialized topics, we would lose the wider perspective. Our focus lies on remaining fully aware of developments in other disciplines and making them available to architecture. We see ourselves as technology scouts for architects.

Over the years we, in the department, have found that communicating certain technologies to architects has often been difficult, even if those technologies are important in further developing the profession. The systematic dealing with CAD systems is one such example. CAD is not only used by architecture students to produce dazzling presentations. However, concepts used in the organization of architectural drawings such as layers, classes, or blocks, or semantic models, like the international foundation classes, are still mostly beyond the scope of their interests.

但是当自由形式的外观成为一种时尚，并且传统制图方式无法便利地设计时，我们能够使诸如参数化CAD这类具有挑战性的系统更加有吸引力，外观项目就吸引一大批学生。量身定做的程序设计带来了巨大的灵活性。所以，建筑师应该学习怎样编程吗？很明显的是，多年以来很少有学生能够接受对他们来说很复杂的、非常奇怪的思维方式。但是当编程生成的构造只能通过CNC(电脑数字控制)设备生成时，建筑学的学生们意识到他们确实需要学习编程，这样才能建造出手工无法建造的构造。通过将数字技术运用到这些现实生活的实例，我们使超过三分之二的学生在设计和规划建筑项目时不仅仅使用普通的CAD应用系统。在与其他建筑类院校作比较时，我们为ETH的成果感到自豪。我们能够使学生们理解到一种可能性，那就是信息技术给他们提供了一种设计依据。

We were, however, able to make challenging systems like parametric CAD attractive to a large number of students through our facade projects, since free-form facades are in vogue and cannot easily be designed using traditional drawing methods. Custom programming leads to an even greater flexibility. So, should architects learn how to program? Previously, for many years, only few students engaged with the complicated and—for them—often quite strange thought patterns that were required. But the programmed structures that could be created using only CNC machines made architecture students realize that they should indeed learn to program, so that they could build structures that could not be manually produced. By tying information technology to these real-world examples, we were able to genuinely involve over two thirds of the students in designing and formulating architectural projects in ways that went beyond the normal use of CAD systems. In comparison to other schools of architecture, this is a result of which we, at the ETH, are very proud. We are able to equip students with a good basic understanding of the possibilites that information technology offers their discipline.

但是，在建筑学中电脑在实践中的重要的运用依然受到了强大的阻碍。我们指的正是无处不在的电脑运用、互联网、转化的建筑学领域、智能屋。不幸的是，由于必要的技术是费力的、混杂的，便相应地出现了一块绊脚石。硬件的生成必须有所需的传感器和调节器；这些硬件必须按照机器的水准编程。进一步地说，单个建筑街区必须被当做一个系统来安装，在这个系统中它们互相联系（比如用远程控制）。

However, an important application of computers in architecture is still being strongly resisted in architectural practice. What we refer to is Ubiquitous Computing, the Internet of Things, or, translated to the field of architecture, Smart Homes. Unfortunately, there is a relatively large stumbling block, since the necessary technology is demanding and hybrid. Hardware must be built with the required sensors and actuators; this hardware has to be programmed at the level of the machines. The individual building blocks, furthermore, must be installed as a system in

这个世界是一个巨大的超级市场！电子建筑模块，我们实验室的成果：太阳能电池、GPS系统、SD卡读卡器、手机、锂电池、Zigbee和蓝牙模块、处理机、色彩展示、二氧化碳感应器、运动感应器、加速度器、触摸感应器、USB接口一点点、一片片地被焊接、插入、缝合起来。

The world is one huge supermarket! Electronic building blocks, the results of our labor: solar cells, GPS systems, SD-card readers, mobile telephones, lithium batteries, Zigbee and Bluetooth wireless modules, processors, colour displays, CO_2 sensors, motion detectors, accelerometers, touch sensors, USB interfaces, bits and pieces to be soldered, plugged in, and sewn in, and much more.

它要求基于互联网的用户界面的实施与互联网的综合化，以及其他更多的措施。它要求许多在不同层面上的技术手段以保证这一系统能够工作，而不仅仅只是一个不错的主意。

幸运的是，现实的世界就如一个巨大的超级市场，并且有许多人愿意用分布式电子系统运作项目。我们已经结合了编程过程平台，它大多被我们使用于物理运算发展。结合 Arduino 系统，我们可以将需要的软件像一个工具包一样组合起来：微控制器、电池、Zigbee 远程控制模块、太阳能电池、运动探测器、高度器、幅度检测雷达、电机控制系统以及更多其他仪器。所有这些可以被置于几个立方厘米的范围内。通过配线，功能相依性可以在硬件上建立起来，并且在网络上通行。CAAD 学科在这些互联网社区里扮演了一个积极的角色。我们的学生进行了一系列有趣的实验，揭示了未来研究的康庄大道。比如说，现在他们可以将电子感应器和"可触媒介"缝进他们的衣服里。另外一组令人印象深刻地联系了物质世界和互联网世界：当它们的主人在 Facebook 上作为朋友联系起来的时候，两个橡胶球靠近对方闪烁和振动。对于一个之前从未学过编程的建筑学学生的选修课来说，这已经相当不错了！

which they are connected, for example, by remote control. It requires the implementation of a web-based user interface as well as an integration with the Internet, and much more besides. It requires skills at many different levels to enable such a system to work rather than remain just a good idea.

Luckily, the world at times is like a huge supermarket, and there are many people who would willingly take on projects with distributed electronics. We have coupled the Processing programming platform, much used by ourselves with the Wiring development environment. Along with the accompanying Arduino system, we can assemble the required hardware like a construction kit: microcontrollers, batteries, the Zigbee remote control module, solar cells, movement detectors, altitude meters, range-monitoring radar, motor control systems and much more. All this can be built on a scale of a few cubic centimeters. Using wiring, functional dependencies can be set up on the hardware and made available across the Internet. The department of CAAD plays an active role in such communities on the Web. Our students carried out a raft of interesting experiments that have unveiled further avenues of investigation. They can now, for instance, sew the electronics for sensors and 'tangible media' into their clothing. Another team has impressively linked the physical world with the world of the Web: Two rubber balls twinkle and vibrate when they come into proximity with one another if their owners are linked as friends on Facebook. Not bad for an elective by architecture students who had never programmed before!

时　　间：2008—2009

参与者：Christoph Wartmann

合作者：Lucerne University of Applied Sciences and Arts

像鸟儿一样飞翔 Flying Like Birds

作为一个技术侦察员，你一定要有耐心。我们知道无处不在的电脑运用或者互联网这些主题对建筑师来说将变得重要。因此我们参与了相关的研究项目，希望为建筑学领域到电子信息技术的专家提供他们感兴趣的应用，研发远程控制的超微电脑。这些共享项目的成果是值得思考的。两年之后我们已经研发了许多具有经济利用潜力的使用脚本，但硬件和软件却与我们设想的不同：我们用一个如同雪茄盒子大小的容器来测量和报告温度数据，它的电容量最多只能供其运作3个小时——这不适用于一项最少需要200个这样的单元体（也称为节点）的项目。

由于这个原因，我们将这个由专家同事制造的"节点"按照我们的需求加以运用。一年之后，三套装置准备就绪，虽然重新调整了"雪茄盒"的尺寸，它们的电池寿命还是不能够满足我们的需要。

改进了多次且经过了多次会议讨论的向国外采购配件的战术失败了，这个挫败可能鼓励了我们，让我们进入了不属于我们领域的领域。我们的需求是简单化和模块化的。我们获益于互联网，就像它今天的获益方式：在这个例子中，超级市场

As a technology scout, you have to be patient. We knew the themes of Ubiquitous Computing or the Internet of Things would become important for architects. Therefore, we participated in related research projects, hoping to offer interesting applications from the field of architecture to the electronics and information technology specialists developing remote-controlled ultra-micro computers. The results of these shared projects were sobering. After two years, we had developed many excellent usage scenarios with economic potential, but the hardware and software was different from what we had imagined: We were confronted with containers the size of a cigar box to measure and report temperature, its massive batteries depleting after only three hours' use—not quite suitable for an application that requires at least 200 of these units, or so-called nodes.

For this reason, we had the nodes built by specialist colleagues exactly to our requirements. After a year, three devices were ready and although now resized to the size of cigarette boxes, their battery life still showed similar limits. They were still not suited to our needs.

The outsourcing strategy had failed, and the frustration, which had developed over time and throughout the many meetings, may have contributed to encouraging us to enter a field not belonging to our own discipline. Accordingly, our requirements were simplicity and modularity. And here as well, we profited from networking as it is

的隐喻很好地形容了这个情形。六个月之内，我们就装配了一套装配系统，这个系统使我们可以在一个星期之内建造出极为高效的火柴盒大小的原件，这正好符合我们的需求。

这个方法产生了许多成功的故事：我们的一位同事在闲暇时间外出进行滑翔伞运动，他利用我们的信号控制技术调节他的滑翔机。为了做到这一点，他在机翼罩盖上沿途放置了10个信号传感器。每一个传感器内都含有一个2.4GHZ传输单元的微型电脑、一个气压计、一个3D加速传感器，以及一个电池组和太阳能块，这些东西被放置在一个容量为20毫米x20毫米x3毫米，仅重8克的小盒子内（不算电池的重量）。因为足够小和轻，因此它可以被直接缝在机翼中。它将数据传输到飞行员肘部的接收器上。接收器由一个微型控制器、一个信号发射器以及相机中的SD卡和手机显示器组成。所有的这些被置于一个用3D打印技术产生的盒子。加之GPS系统，飞行员的手套也被连接起来。飞行员手套的指尖部分装备了如同移动电话中的微型的震动单元。每个手指因此能够直接与滑翔伞罩盖上的传感器相连接，感知机翼受到的压力和加速度。这与鸟翼末端的羽毛非常相似。

经过了长达四年对第三方专家提供物资的无用等待，我们自己将这个技术的可用性转化为了现实，并且

practiced today: In this case, too, the metaphor of the supermarket describes the situation quite well. Within six months, we had assembled a construction system with which we were able, within a week, to construct extremely capable, matchbox-sized elements, tailored exactly to our needs.

There are more success stories resulting from this approach: One of our colleagues at the chair goes paragliding in his spare time and used our radio-control technology to tune his glider. To do this, he placed ten radio sensors along the wing of the canopy. Each consisted of a small computer with a 2.4GHz transmitting module, a barometer, a 3-D acceleration sensor, and a battery or solar cell, all within a volume of 20 x 20 x 3 mm and with the weight of only 8 g (without battery). Small and light enough, therefore, that they could be sewn directly into the wing. They transmitted data to a receiver on the pilot's wrist. The receiver was made from a micro-controller and a radio transmitter, along with an SD card from a digital camera and the display from a mobile telephone. All this was built into a case which we produced using our 3-D printing. In addition to the GPS system, the gloves of the pilot are connected as well. The fingertips of the pilot's gloves were equipped with small vibrating units much like those known from mobile telephones. Each finger was therefore directly connected to a sensor in the paraglider's canopy and could feel the pressure and acceleration from the wing transmitted using vibrations. This setup resembles much the leading feathers at the end of a bird's wing.

After four years of waiting in vain for third-party experts to come up with the goods, we made the technology available by ourselves, and

滑翔伞的装备：早期版本的发射模块（上图），带有 GPS 和记录仪的接收模块（中图），指尖带有内置压电振动器的手套。

The equipment for the paraglider: the transmitting module in an earlier version (above), the receiver module with GPS and recorder (middle), and the gloves with the inbuilt piezo-vibrators for the fingertips.

瑞士阿尔卑斯山上正在使用中的系统。

The system in use in the Swiss Alps.

很快惊喜地获得了回报。就在公布我们的成果之后的短时间内，许多知名的滑翔机生产厂商想让我们加入生产。很快我们决定和在 Thun（瑞士）的一家公司合作，他们的产品最近获得了滑翔机世界杯的冠军。我们目前正在合作使将来的滑翔机更安全、速度更快。

这个例子证明了，首先，跨学科项目需要适当的理解力和在学科壁垒间"转化"的能力；第二，它们有着惊人的伪装。在这种情况下，一个试图了解当今技术、并使其成为日常可用技术的建筑师，他的心态是绝对不能低估的！

were surprisingly quickly rewarded. Shortly after the first publication of our findings, many well-known hang-glider manufacturers approached us to participate. The decision finally went to a firm from Thun (Switzerland), whose products had recently won the hang-gliding world championship and world cup. We are now cooperating to make tomorrow's hang gliders safer and faster.

This example shows that, first of all, successful transdisciplinary projects require a proper understanding and a capability for 'translating' across disciplinary borders, and, secondly, they come in surprising guises. The mindset of architects who are trying to understand present technologies and to make them available for everyday use should not, in this case, be underestimated!

时　　间: 2008—2009

参与者: Benjamin Dillenburger und ca. 80 Studenten

合作者: Prof. Matthias Castorph, Universität Kaiserslautern (D)

建筑谷歌 Architectural Google

你可曾想过只需轻击一下鼠标就能解决一个建筑问题？或者只需基地平面就能得到一个建筑平面图？现在有一种专门为建筑设计服务和短信一样快捷的全自动功能可以做到这一切。

对于建筑师和城市规划人员来说，工作的时候参考文献和以往的经验早已成为习惯。先前的工作经验是一笔极有价值的财富。用类比过去经验的方法工作能更好地阐述和沟通自己的想法以及确定一些重要的细节，或者，它有时也可作为后续设计的起点。不过目前这样的经验还无法在程序设计中做到标准化地存储和提取。在最理想的状态之下，建筑师可以从其以往的个人经验中得出合适的方案，依据之前收集的实际项目绘图，或者去翻找标准设计的目录。这样的搜索不是主观性太强、选择太少，就是完全拘泥于死板的指令。搜索要遵循一定的原则或者标准形式，搜索必须具有一定的法则，样本才能被检索。因此，我们的项目在描述范畴有一个突破，任意一个设计不管怎样都能得到一个标准化的形式。这个过程就不再局限于建筑平面图和草图的存储与提取。正因为这样，建筑工作室的工作也不会仅仅局限于建筑师个人的经验。

What would it be like if a computer could give an architect possible suggestions for solving a certain architectural problem with the press of a button? Or suggest a suitable floor plan from a site drawing? An 'auto-complete function' for architecture, say, just like for SMS messaging.

Working from and with references is professional routine for architects and urban planners. Earlier projects constitute a valuable body of experience. Drawing on analogies can serve the illustration and communication of ideas and the determination of specific characteristics, or it can serve as a starting point for further planning. Until now, architectonic experience has not been amenable to standardization or storage and retrieval. In the best-case scenario, the architect will remember a suitable design solution from personal experience, will be able to draw upon collected projects from within the practice or go back to catalogs of standard plans. The search follows either a subjective and limited selection or a rigid ordering principle. The collections are organized according to certain formalisms or standard forms, and the examples can only be found if the search can be described in the forms. Above all, this constitutes an abrupt change of the scope of the description; the plan will, one way or another, get shoehorned into a standard form. This procedure is, therefore, far too constricting for the storage and retrieval of architectural plans and drawings. Because of this, architectural offices are far too dependent upon the personal—and limited—experiences of the people working there.

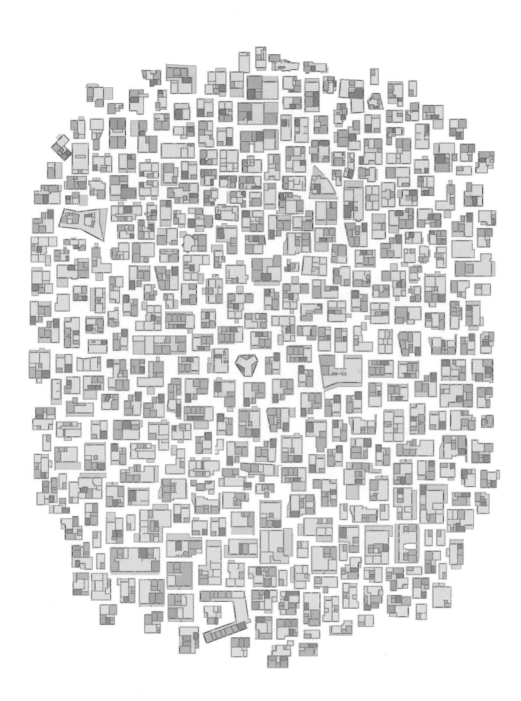

目录索引底层平面图。

Indexed ground plans of the catalog represented as scheme.

#121

#122

索引化底层平面图切割成硬纸板模型，堆积起来成为一幢"高层建筑"，左侧鸟瞰图，以及此方向的侧透视图。

The indexed ground plans as cut-out cardboard models, stacked to make a 'high-rise,' left-hand in bird'side in profile perspective.

相比之下，用大型的搜索引擎（比如谷歌）进行搜索和浏览网页，往往能有丰硕的成果，即使在使用过程中会受到一些其他信息的干扰。不过，谷歌的优势还是很明显的。你无需在搜索时使用一些之前设定好的搜索模式以保证搜索结果的有序性，因为谷歌搜索的是一切包含搜索字符的文章。所以，那些之前设定好的搜索模式也就随之退出了历史舞台。即使人们一直在呼吁改进搜索引擎，但若是有一天没有了搜索引擎，真是无法想象我们的生活会变成什么样子。人们更青睐谷歌是因为谷歌的搜索结果是根据点击率的多少来排行的。谷歌在进行搜索时不考虑搜索内容的意义，也不会通过分析字面意义来衡量信息的重要性。信息的意义全都留给用户自己去评判。当搜索结果包含一系列基于点击率排行的文章索引时，使用者常常会觉得搜索引擎似乎能"理解"他们所提出的问题。显然，这样得出的搜索结果会比字面对应的搜索要来得好得多。在我们看来，这是人机合作历程上的一个重大突破。

诚然，设计草图目前还不能像文字一样被直接搜索。我们和Kaiserslautern 大学的 Matthias Castorph 教授合作制作了一项包含1 000个底层平面图及其文字说明的目录。这些平面图的几何布

In comparison, searching and browsing the Internet using one of the larger search engines, such as Google, clearly yields a wealth of results, even if the profusion of irrelevant hits may sometimes be annoying. Still, the advantage remains with Google & Co. that you don't have to use a predefined form for searching that would demand the aforementioned change in the scope of description; rather, they look for articles containing the text fragments entered. The structural limitations of predefined search forms—with their specific notions about the structure of the possible contents—simply disappear. And despite all the moaning and calls for improvement, it would simply be unthinkable to do without these search engines, which we use on a daily basis! Google, in particular, only gained its pre-eminence because, with its link- and usage-based page ranking system, it did away with the need to understand the meaning or to evaluate the information content at the explicit level of the text. The assignment of meaning was left to the user—or to the community of users—of the search result. The user gets the impression that the search engine somehow 'understands' the question posed, when in fact it only returns a list based on the usage and linking of indexed articles. Naturally, this gives better results than the textual-analysis algorithms that are known today—in our opinion, a groundbreaking concept for the cooperation of man and machine.

Admittedly, there is still no convincing mechanism by which drawings can be searched on the Internet as easily as can text. Together with Professor Matthias Castorph, from the University of Kaiserslautern, we have put together a catalog containing, as of this writing, 1000 ground plans

局、面积、地形全都用文字加注，如此一来用户使用谷歌一类的搜索引擎时就能够搜索到这些图了。我们现在可以直接在谷歌中输入一个平面图中的字段，就会得到许多包含该字段或相似字段的平面图。倘若我们向所有的建筑师提供一个基于所有瑞士建筑的平面图的索引数据库，或是像我们在瑞士合理化建设中心项目中做的那样，会有怎样的结果呢？在新的工作方式、新的教育与学习方式、新的经营模式出现以后，新的建筑毫无疑问也会随之出现。

for dwellings. The geometry, dimensions and topology of these ground plans are translated into words and sentences, and are therefore made searchable by engines such as Google. And with no small success! We can now enter a fragment of a ground plan as a search term in Google, and entire ground plans containing that fragment—or similar fragments—will be returned as search results. What would happen if we used this indexing machine upon the entire building stock of Switzerland, and if we—as in project CRB online—make it available to every architect? New methods of working, new methods of teaching and learning, new business models, and new architectures would, undoubtedly, result.

时　间：2006

参与者：Markus Braach, Oliver Fritz, Benjamin Dillenburger, Alexander Lehnerer, Steffen Lemmerzahl

合作者：Architekturbüro Jaschek (Stuttgart, D); SLIK Architekten GmbH (Zürich, CH); group8 (Geneva, CH)

制造建筑的机器人 An Automaton for Making Architecture

观望近几年来数字技术的狂飙式发展，技术本身被高估，使我们把幻想寄托在它身上，也使我们问类似于这样的问题："计算的局限性在哪？电脑可以变得聪明吗？可能出现一个能制造建筑的机器人吗？"

我们认为这些问题给出了错误的方向。它们很可能并没有抓住技术在我们生活之中的真实意义——特别是信息技术。并且它们使我们把注意力从真正有趣的任务上转移开来。

理想主义的简化论者坚信人类与（数字）机器间的对抗，甚至是冲突——或者说是一种对于它们之间永远无法协调的信仰——是没有任何结果的。正是具体的项目在我们的环境中规划出目标，我们相应地开发出机器去实现了这些目标。当技术从我们的思考方式中分离出来，变得类似独立的时候，我们相信人类和技术之间的关系——特别是随着后者变得越来越强大——将会变得复杂。如果实际工具失去了与它的特殊用途之间的关系——哪怕这只是在谈论开始时的例子，错误的问题也会因此而出现。另外，棘手的、几乎是"不现实"的问题很容易用消极的方式回答，以至于将我们的注意力从问题的核心转移开。

Looking into the stormy development of digital technology over recent years, it is tempting to overestimate technology itself, to subordinate our projecting fantasy to it, and to ask questions such as: 'What are the limits of computation? Can a computer be intelligent? Will there ever be an automaton for making architecture?'

We believe these questions point us in the wrong direction. They probably do not capture the real meaning of technology—and especially information technology—in our lives. And they divert attention from the really interesting tasks.

The idealistically reductionist claim of antagonism, even of conflict, between man and (digital) machine—or the belief that they will never be fully reconciled—leads us nowhere. There are only concrete projects, formulated as goals in our environment, and there are machines that we correspondingly program to achieve these goals. We believe the relationship between humans and technology—especially as the latter becomes ever more powerful—becomes difficult only when technology is abstracted from our ways of thinking and becomes quasi-independent. If consequently the link between the actual apparatus and its specific function is lost—even if this were only the case in an initial discourse—the wrong questions will come to the fore. Additionally, the thorny and almost 'impractical' questions can easily be answered in the negative, and so divert attention from the core of the problem.

但是这不应该阻止我们将我们的思想回到一个熟悉的人物身上：你可能会记得 ELIZA，一个由 Joseph Weizenbaum 于 1966 年设计出的人造精神病医生。ELIZA 用简单的算法向她的"病人"提问，于是产生了以下类型的"对话"：

病人："我和我父亲之间出现了一些问题。"

ELIZA："再告诉我一些关于你家庭的事！"

虽然参与者们知道他们正面对的是一个软件程序而非精神病医生，大多数人仍然感觉自己被 ELIZA 理解了。人们并没有认为机器会在不理解他们想法的情况下用机器的方式将单词组合起来。这是一个多么有趣的现象！

还有更多的例子，你可能还知道深蓝，1966 年由 IBM 公司开发的国际象棋计算机，打败了加里·卡斯帕罗夫——当时的国际象棋世界冠军。这次胜利并不是由于安装了书本上的国际象棋专业知识，而仅仅是纯粹通过运算速度来运用极其简单的国际象棋法则。速度迎战智力。

但是一个建筑"设计机器"应该是什么样的呢？我们一直在进行一项实验，去创造一个类似于建筑布局合成器的东西。在我们的系统里，取代生成器、振荡器、调制器、放大器和过滤器，我们的合成器运用路径长度、邻里、日照、通风、基地坐标系、楼层高度、轴向布局等等更多的规则。这种机器可以根据它最初的设置参数

Still, this should not stop us from casting our minds back to an old acquaintance: You may remember ELIZA, the artificial psychotherapist developed in 1966 by Joseph Weizenbaum. ELIZA reformulates questions from her 'patients' using simple algorithms, so that 'dialogs' of this sort were produced:

Patient: 'I have a problem with my father.'

ELIZA: 'Tell me more about your family!'

Although the participants knew they were dealing with a software program and not a psychiatrist, the majority felt understood by ELIZA. People do not assume that machines could put words together in a mechanical fashion without understanding their meaning. That is an interesting phenomenon!

By further example, you my also know Deep Blue, the chess computer developed by IBM, which, in 1996, was able to defeat Gary Kasparov—at that time the reigning chess world champion. This victory was not won by programming in expert knowledge from chess books, but simply through pure computing speed acting upon the extremely simple rules of chess. Speed versus intelligence.

But what might a 'design machine' for architecture look like? We have been working on an experiment to develop something like a synthesizer for architectural layouts. In our system, instead of generators, oscillators, modulators, amplifiers, and filters, our synthesizer has rules for the lengths of paths, for neighborhoods, insolation, ventilation, site axes, story heights, axial layouts, and far more

与多种初始参数相符的一个建筑设计的不同变体(从上到下)的逐层表现(从左到右)。在这个案例中,不同颜色表示 60 000 平方米的办公建筑中单个空间的不同使用功能。

Story-by-story representation (from left to right) of different variants of a building design (from top to bottom) corresponding to diverse initial parameters. In this case, colors show the various uses of individual spaces of a 60,000 m² office building.

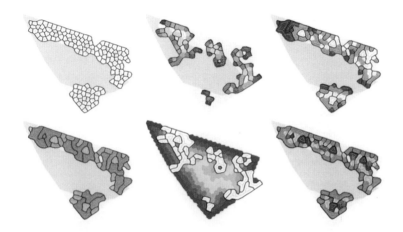

一个总平面布局上的各个单独空间由不同的品质（如视线、路径长度、日照、垂直交通）计算而得出不同形态，以便于评估这个布局以及与其他布局相比较。

The individual spaces of a layout are calculated for various configurations across several qualities (e.g., view, length of paths, insolation, provision of vertical accessibility) in order to evaluate the layout and compare it to others.

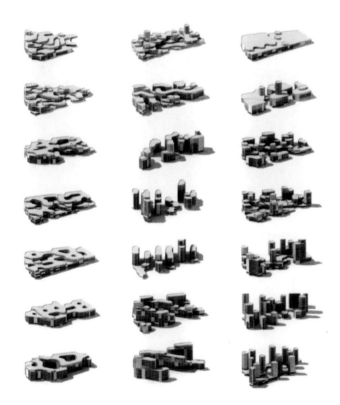

当运用相同的空间分配平面时，不同的建筑布局生成，取决于对不同品质的权衡（如视线、路径长度、日照、垂直交通；这些布局以 3D 模型的方式展示在这里。

While using the same space allocation plan, different architectural layouts are generated, dependent on the weighting of the various qualities (e.g., view, length of paths, insolation, provision of vertical accessibility); the layouts are 3-D-models here.

全自动生成结果的逼真表现。

Photorealistic representation of the fully automatically generated result.

应对复杂的空间结构并生成一个建筑学上可行的总平面，生成的结果可以被逼真地表现出来，并且参加了建筑竞赛。到目前为止，我们获得过第四名，但是很不幸还没有赢得过竞赛。因此，建筑设计可以自动化吗？当然可以！但是这是个错误的问题。建筑学万岁！

besides. Such a machine is capable of taking a complex spatial structure and producing an architecturally feasible site plan, according to its initial programming parameters. The results are rendered photo-realistically and entered into architectural competitions. Thus far, we have been able to achieve a fourth placing, but unfortunately no wins. Can architectural design, therefore, actually be automated? Of course! But it's the wrong question. Long live architecture!

时　间：始于 2008

参与者：Matthias Bernhard, Philipp Dohmen, Steffen Lemmerzahl

合作者：digitales bauen GmbH (Karlsruhe, D); Halter Unternehmungen (Zürich, CH); ERNE AG Holzbau (Laufenburg, CH); Aepli Metallbau (Gossau, CH)

在施工现场 On Site

纵观我们的建筑师同行们对这个领域所做的贡献，总会找到理论或者技术实验性的例子。对于建造工程师来说，形式有些不同，但是结果近似。很少有项目用实际参照来构想他们的方案，更少有可行的商业模式。我们部门创建于 2000 年，目标是建构信息技术的理论和模型并成功应用于建筑，在实践中调整，而不是在电脑上基于假设来发展实验性建筑模型。这也许让人非常惊讶，但是这种务实的研究态度在这个领域中并不常见。

建造工业至少在西方发达国家是高产值的产物。它的产值比食品、药品、工程、保险、保健和旅游产业更高，但是却很大程度上被忽视了。更不幸的是，建造产业没有很好的口碑和前景。这可能缘于它很大程度上由工匠技艺主导，只有很少部分的学术和工业活动。因此，在这点上建造有很大的发展前景。

在现实世界的经济下价值创造的多种可能性是我们研究的重要方面。但是，找到这些机会通常很难。和我们的同行一样，我们最

Browsing through the contributions of our architect colleagues to conferences relevant for our field, one finds almost exclusively theoretical and technical-experimental examples. For building engineers, the situation differs a little, yet not the results. Few research projects formulate their theme with a practical reference, and fewer still reflect possible business models. Our department was established in the year 2000, with the stated aim of formulating theories and models of information technology that could be successfully implemented in architecture and verified in practice, not of developing experimental architectural models in the computer on an 'as if' basis. It may be surprising, but this pragmatic thrust of our research is not often found in our field.

The construction industry is—at least in Western industrialized nations—the industry with the highest gross value creation. It is higher than that for the food, pharmaceutical, engineering, insurance, health, or tourism industries, but this aspect goes largely unnoticed. Unfortunately, the construction sector does not have a very good reputation and has low prominence. This may be due to the fact that in large parts it remains dominated by artisanal skills, with only pockets of academic or even industrial activity. For this reason, it bears an enormous potential for further development.

Possible opportunities for value creation in the real-world economy are an important aspect of our research. However, finding these opportunities is often difficult. Like many of our colleagues, we

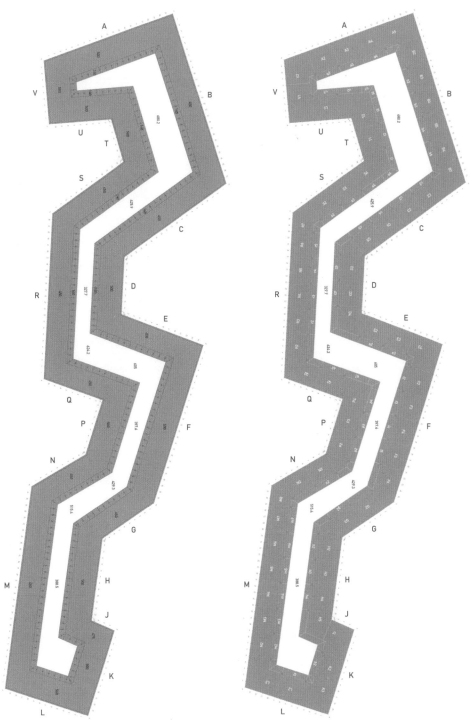

米制单位系统和建造系统都是使用数字化方法。

The metric system and the construction system of the building using the 'Building Digital' method, Karlsruhe (D).

开始专注于建筑的设计部分。我们创造了新的几何形状、新的设计语言，我们发展了新的规划过程和建造方式。对于这些项目，我们从市场或者更好的建筑实践中获利。因此，随着价值创造的发展，我们在整个建筑每英里的投资中的贡献仍然很小。从经济学的角度来看，我们的技术实践在某种程度上比我们作为复杂建筑规划咨询者做得更好。这个商业模型是我们的副产品之一，也是Design2Production公司所追求的。这个公司已经在建造项目中增加了个位数百分比的价值。

但是，如果我们成功地从建造形式和建筑管理中抽象出标准化和工业化的日常生产建造过程的方法，而且很大程度上从它们的几何形式中独立出来，那么将会增加30%的潜在价值。这不会在某些针对性市场中发生，但会在我们的国有经济这个最大市场中产生。诚然，这将是建筑师所面临的设计过程的最大挑战，要考虑到每一个可能的形式，因而它们自身不会存在形式。几乎我们所有的同行都利用新技术发展不同寻常的新形式，来以此提高他们自身的形象。

我将讨论一个从当地文化中抽象出来的具有潜力的项目。这个项目由三家大型瑞士公司共有，一家是综合建造企业，其他两家专长于立面建造。他们针对建筑和

initially focused on supporting the design part in architecture. We have created new geometries, new design languages, we have developed new planning processes and new construction methods. For these projects, we were paid from the marketing budget or—in the best case scenario—from the architecture practice's fees. Therefore, as far as value creation goes, our contribution has remained financially insignificant in around the per-mille range of the total investment for the building. From an economic perspective, the implementation of our technologies has performed somewhat better when we have worked as planning consultants on the production of geometrically complex constructions. This is a business model that one of our spin-off projects, the firm Design2Production, has pursued. The firm has added value in single-figure percentages to the building projects in which it has taken part.

However, if we successfully abstract from the built forms and architecture and manage to standardize and industrialize the building process of everyday buildings, largely independently of their geometric form, this increases to a potential 30% value add. This would not take place in some niche market, but in the biggest market in our national economies. Admittedly, it is an extraordinary challenge for architects to design processes that take into account every conceivable form, and thus have no form themselves. Almost all our colleagues use these new technologies to develop unusual visual forms, and in so doing, raise their own profiles.

I would like to talk here of a prominent project that was abstracted from the peculiarities of the local and cultural vernacular. This project, shared between three large-scale Swiss companies—one general construction concern and two facade

建造过程建立的正式方案，能够概括瑞士在 2009 年 80% 建筑所用的一种常见结构的标准。直到现在，工业化建筑仍然从属于建筑的一个狭窄系统（"首先是系统，然后才是建筑"）。在这个项目中，系统是针对建筑提出的（首先是建筑，然后才是系统）。但是，我们严格区分形式和结构领域，并且在结构允许的范围内采用灵活的形式。在这个项目中，这个实验性建筑采用三个建造步骤模型：

首先这个建筑外壳大部分是采用很难运输的材料，在现场施工，比如现浇混凝土。任何形式的复杂性都从建筑外壳中舍去，没有复杂的体块，没有耐用性的要求，采用和建筑技术总是匹配的接口。这使得当地公司在没有任何特殊指导和风险的情况下生产建筑外壳。

这个建筑较为复杂的技术方面（卫生设施不在考虑之中）都集中在立面上。立面元素的生产就像汽车一样使用了现代工业技术。我们区分技术核心（好比汽车的发动机）和建筑外壳（汽车车身）。技术核心包括模块概念，涉及如气候控制、通风、光照、遮阳、电力、数据、信息、安全、声控等等。其中最重要的技术核心就是每个都被打包进入立面模块中，能够被自由设计。有了这些智能自动化的立面模块，我们能够针对建筑设计的需求来给建筑"傻瓜"式地赋予表皮。用倾角切削技术来生成

specialists—came up with formal descriptions of building and construction processes that could map up to 80% of the buildings built in Switzerland in 2009 within one common structure. While until now, industrialized building has meant subordinating the architecture to a narrow system ('first the system, then the architecture'). In this project, however, the system was organized according to the architecture ('first the architecture, then the system'). However, we strictly separate the formal and the structural domains and obtain a formal freedom through a remarkable structural stringency. In this project, the pilot buildings are constructed according to the following three-stage model:

Firstly, the building shell—and with it a large proportion of the difficult-to-transport materials—is constructed on site, for example, in concrete. Any kind of complexity is being removed from the building shell: no complicated massing, no demands placed upon the tolerances, always the same interfaces to the building technology, and so on. This makes it possible to have the building shell erected by local firms without any special directives or implying any particular risks.

The more complex technological aspects of the building (the sanitary installations are excluded from this consideration) were clustered into the facades. The facade elements are produced with modern industrial techniques, rather like cars. We distinguish the technological kernel (comparable to a car's motor) from the architectural hull (the body work). The technological kernel consists of a modular concept of decentralized, intelligent building technologies for all aspects, such as climate control, ventilation, lighting, solar protection, electrics, data, communication, security, audio, and far more. This mostly identical technological kernel is being individually packed into facade modules, that can be freely designed. With these technologically autonomous facade modules we clothe the 'dumb'

外壳。这些立面模块用软件相连，因此整个建筑正如被预想的那样是一个高水平系统。

在第三阶段，进一步建造开始实施，依然是当地的公司在现场施工。这个阶段，需要考虑业主的个人需要。幸运的是，这个问题已经在技术层面上解决了，因为智能化模块能够通过软件适应很大范围内的个人需要。考虑到传统元素，建筑在当地传统范围内，可能更加手工业化。而这里，因为在建筑服务工程中标准化生产，建筑可能会进一步改变。

这个过程意味着立面元素可以在工厂中生产，从而明显增加了它们的价值。因为这个过程很大程度上和实际建筑分开，所以我们期待大规模生产和大生产力。这些元素得以发展并细致考虑到高效、舒适、安全的需求。质量水平因此都能达到很高的程度，而以今天的手动生产方式这是不可能的。并且，高质量的建造可以成为出口商品。

对这个项目的兴趣在于在这条精选的路径上，没有特别的科技阻碍。相反，问题源自已经建立的规划和建造过程、投标过程和各种商业责任。许多股东需要承担对新工作方式的责任，需要重新定义他们的角色。从根本上讲，我们面对一个困难的说教任务，面向这些每个人都用自己想要的方式竞争的大大小小的部门。

shell according to the architectural design, and mount the shell with cutting-edge technologies. The facade modules are connected by software so that the whole building works, as planned, as a high-level system.

In the third step, further construction is carried out again by local firms on site. This stage requires consideration of the individual needs of the client. Fortunately, however, it is already relieved of any technological complexity, since the intelligent technological modules can be customized to a wide range of individual needs through software. Concerning their traditional elements, the buildings remain within the remit of the local, and perhaps more artisanal building industry. Here, the buildings may differentiate further, in spite of or just because they are standardized on the level of the building services engineering.

This procedure means that the facade elements can be produced in a factory, a process that significantly increases their added value. Since these processes are to a large extent independent of the actual architecture, we expect large lot sizes and, correspondingly, a high productivity. The elements are able to be carefully developed further to take into account the increasing demands of efficiency, comfort, and security. A level of quality can therefore be achieved that is simply not possible with today's manual production methods. Furthermore, top-quality building construction can become an export commodity.

What was interesting about this project was that on this chosen route, there are no particular technical obstacles. Rather, problems arise from the established planning and construction processes, the tendering procedure and the responsibilities of the various trades. The many stakeholders need to gain trust for new working methods and to redefine their roles. Primarily, we are faced with a difficult didactic task in a small-to-medium sized sector in which everyone competes in any way he wants.

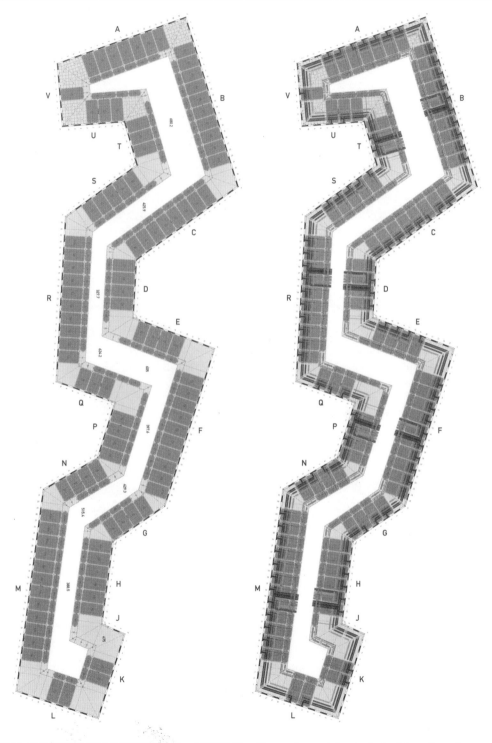

该技术系统针对于供应和浪费问题,同样针对于利用建筑数字技术创造和处理的基础模块。

The system of the technical locations for supply and waste disposal, as well as the modules of the technical infrastructure as created by the 'Building Digital' method, Karlsruhe (D).

时　　间：2004

参与者：Wilfried Beck, Balz Halter u.v.m.

合作者：aizo ag (Wetzlar, D); digitalSTROM alliance (Zürich, CH)

电子智能 Electrical Intelligence

最显著的基础项目是从我们部门中心研究发展而来，我们称它为digital-STROM。正如名字暗示的，这个项目证明了两个领域以及不同的思考方式的跨越，即电力供应、数字编码信息这些看起来截然不同的因素被应用于设计。

第一个领域是用电流的形式进行的能源供应。在120年里，电已经让"能源到处都在"变为可能。合理分配能源的问题已经被恰到好处地解决。在电气化时代之前，能源的运输是很费力的事情，蜡烛、木头、煤炭必须被运输。在电出现之前的时代，供应可用能源处于古老而脏乱的状态，伴随产生火、灰、烟，这从史前时代开始就很少改变。随着从120年前电力的发展，大量令人眼花缭乱的设备也得以发展，就像使用火一样让人觉得匪夷所思，电灯、门铃、微波炉、冰箱、电钻、电视、电话、咖啡机、抽油烟机、吸尘器、信号灯、机场的传送带、高层中的电梯、大型建筑的中央空调、报警系统、火警、自动取款机、燃油喷射泵、检测怀孕的超声波系统都使用了电能。我们估计全球有3 000亿至5 000亿的电力设备。相比之下，所有全球

The most prominent infrastructure project which originated from within our department centers around what we call digital-STROM. As the name implies, this project demonstrates the crossing over of two areas of life and ways of thinking that are generally seen as being distinct: electrical power supply and digitally encoded information.

The first area was the supply of energy in the form of electric current. For around 120 years electricity has made it possible that 'energy is always there.' The logistical problem of distributing available energy has been resolved elegantly. In the days before electri-fication, the transport of energy was a laborious business: candles, wood, or coal had to be carried around. In those pre-electrical times, the provision of available energy was an archaic and dirty state of affairs, generating fire, ash, soot, and smoke, something that had changed little since prehistoric times. Since the development of electricity 120 years ago, a dizzying variety of electrical equipment has been developed, something that would be unthinkable just using fire: electrical lighting, doorbells, microwave ovens, refrigerators, electric drills, televisions, telephones, coffee machines, extractor hoods, vacuum cleaners, traffic lights, travelators at airports, elevators for high-rise buildings, air conditioning units for large buildings, alarm systems, video surveillance systems, bank ATM machines, fuel-injection pumps for internal combustion motors, ultra-sound scanning systems for monitoring pregnancy, and so on. We estimate around 300 to 500 billion pieces of electrical equipment exist globally.

的电力设备中，有 20 亿至 30 亿的手机。

但是，所有的用电设备都是不智能的消费者。它们被插入插座，然后消耗电能。它们并不能够适应所在的环境，既不能告诉我们它们运行的状况，也不能接受信号或者像一个团体一样一起工作。比如，当你离开家，所有的设备必须拔掉插座，没有一个中心管理所有的设备。再比如，我们的房子中有大量摇控器，因为这些设备并不能彼此交流，因此必须单个设置时间，使用复杂的不同按键。电子设备是能源的简单接受者，能源从几个中心位置被分配。作为一个消费者，我很感激我能得到我所需要的能源。

这是跨越信息技术这个点子的来源。让我们来看看网络，众所周知，网络是由不同的规则支配的。每个参与者都是信息的发送者和接受者。作为终极使用者，我能够发送和接收邮件。通常在网络上有很多的数据，有很多种方式来处理这些数据，全球网络、电邮、谷歌、维基百科、脸书、推特等等。如果电子设备也形成这样一个相似的系统，那么将会是怎样的？

digitalSTROM 一开始就旨在解决这个问题。如果 STROM 意味着"能源到处都在"，它的存在改变了我们每天使用的电子设备的局限性，那么 digital-STROM

The worldwide total of some two to three billion mobile telephones pales in comparison.

However, all these electricity-using devices are 'dumb' consumers. They are plugged into the socket and consume power. They cannot adjust to their environment; they neither inform us of their condition nor are able to receive signals or to work together as a team. When, for instance, you leave your house, every individual device must be switched off—there is no form of 'central locking' for these devices. Another example is the profusion of remote controls in our living rooms; because all these pieces of equipment do not 'talk to' each other, the time has to be set individually for each one, using a bewildering variety of different buttons. Electrical devices are simply receivers for energy; energy is distributed from a few central locations and, as a consumer, I am thankful that I can get enough energy for my needs.

This is where the idea of a crossing-over with the information technology sphere comes in. Let us examine the Internet, where, as is generally known, different rules apply. The individual participants are senders and receivers of data. As an end user, I can send and receive e-mails. There is an abundance of data on the Internet and there are many ideas as to what one can do with all this data: World Wide Web, e-mails, Google, Wikipedia, Facebook, Twitter … What would it be like if electrical devices formed a similar system?

The digitalSTROM initiative concerns itself with answers to exactly this question. If STROM means that 'energy is always there,' and this availability sets the limit for the many electrical devices in our everyday lives, then digitalSTROM means that

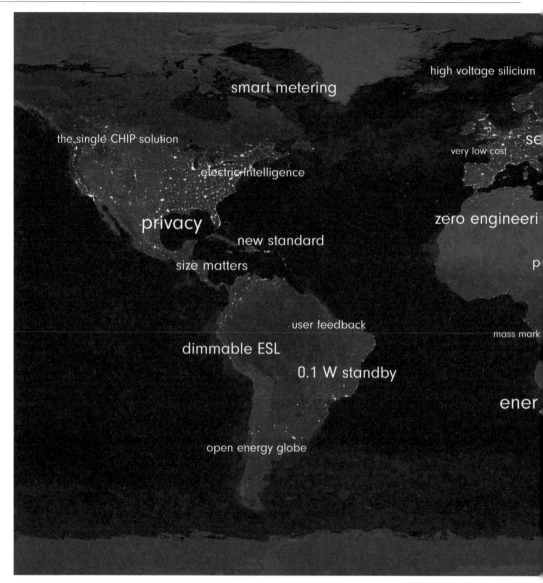

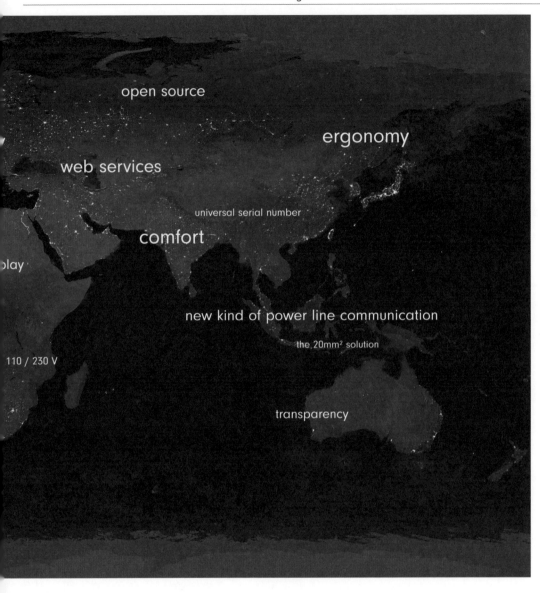

世界在2009年被500亿电子设备所照亮。

The world in 2009 is 'illuminated' by around 500,000,000,000 electric devices.

就意味着"系统到处都在"。在这个系统下,电子设备被植入,它们能够互相交流,就像它们是某种感应单元。这项改变的结果是巨大的。

在这个系统里,电子设备即使是电灯,也不再只是用开关控制的,而是通过能够控制能源状态的设备,或者使用者通过信息化技术与这个设备沟通。这意味着我们和电力设备的关系会从根本上改变。

作为一个使用者,我不会去开或者关一个台灯,这个台灯总是会在线。我能够根据行为的模式或者其他照明设施的组合来控制。取代原来按开关的方式,我可以简单地对房间的电力系统说话,告诉它灯是太亮或者太暗。我可以告诉房间音乐是过大或者继续播放。我可以告诉房间我想要睡觉,或是我听力不好,或是我要去度假,或是从下午4点开始。从今往后,电力会有更高的利用率。电力设备会根据不同的特点和需要一起工作。

从技术的角度来看,使用我们在苏黎世和韦茨拉尔的公司所设计的小型高电压芯片是可行的。这个芯片用主要的线路系统可以建立一个当地网络中心或者信息频道,比之前的体积小100倍,而且比现在市场上的任何替代品少消耗30%的电能。它是第一个能够适应任何设备

the 'system is always there.' In this system, as soon as electrical devices are plugged in, they talk to each other as if they were some kind of responsive units. The consequences of this transformation are enormous.

In this system, an electrical device—even as ordinary as a lamp—is no longer energetically 'controlled' with a light switch; rather, it either controls its own energetic status as an information agent or reacts to information from the user. This means that our relationship to electrical installations will change fundamentally.

As a user, I wouldn't turn the electricity to an individual lamp on or off—the lamp would always be online. I could switch it individually and directly to a particular mode of behavior, or in combination with all the other lighting devices. Instead of flicking an electrical switch to influence each individual lamp, I could simply say to the 'room' electrical system that it is either too bright or too dark for me. I could tell the room that the music is too loud for me, or to continue playing. I could tell the room system that I want to sleep now, that I am hard of hearing, that I'm going off on my holidays, or that from four o'clock pm. Onwards, the electricity will be billed at a higher rate. The individual electrical devices would behave together according to their characteristics and on the basis of various demands.

From a technical perspective, this can be made possible using a small, high-voltage chip that we have developed with a company situated in Wetzlar and Zurich. This chip can set up a local hub or a communications channel to its 'colleagues' over the mains wiring, and is therefore 100 times smaller and uses 30

和插座的产品。而在当前的技术支持下，要做到这一点必须占有更多的空间，消耗更多的电流，散发更多的热量，这会使许多细小的零件融化。

digitalSTROM 是第一个切实可行的解决方法，将全球的 5 000 亿电子设备转化为一个信息技术系统。它为我们的环境打开了新的创新层次，和电、手机或者网络相提并论，它由建筑师、电子专家、信息技术者共同创造。

times less current than any of the alternatives currently on the market. For the first time, it can be built into every device and every plug. With available technologies, this would need too much space, and would create so much heat through increased current consumption that those small cases would melt.

digitalSTROM is the first viable solution to turn the 500 billion electrical devices in the world into an information technology system. It opens a new level for innovation in our environment, on a par with electricity, mobile telephones, or the Internet—and it was invented by architects, electronics experts, and information technologists.

时　间：始于 2005

参与者：Vera Bühlmann, Sebastian Michael u.v.m.

能源过剩 Energy Abundance

建筑师经常寻求更广阔的设计视角，比如勒·柯布西耶、国际现代化建筑大会、Buckminster Fuller、阿基格拉姆学派，或者我的老师 Fritz Haller，他著有《Totale Stadt》。当前这种乌托邦式的理想可以从近、中和远东地区的新城项目中看到。

如果你能让我作为一个设计师制定一个特定的思路，我将不胜感激。目前，能源短缺和气候变化问题在世界范围内正被深入探讨。节约能源的要求不断给我们留下深刻印象，但是我们也许更应该牢记每年到达地球表面的太阳辐射能量比目前人们可使用的还要多大约一万倍。因此，这个问题不是一般概念上的能源短缺问题。很明显，这种困难甚至是一种文化问题，即我们该怎么做才能得到这丰富的资源而在其生产过程中不产生出过多的二氧化碳？

在节约能耗的呼吁中，容易忽略的是我们确实有可行的解决方案。然而，这给我们对能源的思考和处理方式带来了挑战。例如，我想加强关注薄膜光伏技术。这种生产有效能源的技术有以下根本性的新特点：

在 2009 年年中，使用太阳能膜光电技术的成本大约是每峰瓦 50 美分。"峰瓦"（峰值功率）是一个

Architects have often looked for wider perspectives: Le Corbusier and the Congrès International d'Architecture Moderne (CIAM), Buckminster Fuller, Archigram, or my teacher Fritz Haller, with his book 'Totale Stadt' (Total City). The current state of this past utopian ideal can be seen in the Newtown Projects in the Near, Middle, and Far East.

I would be grateful if you would allow me, as an architect, to formulate a particular train of thought. At the moment, the problems of energy shortages and climate change are being intensively discussed worldwide. The need to conserve energy is continually being impressed upon us, so it is probably worth keeping in mind that every year the sun radiates about 10,000 times more energy onto the surface of the Earth than humanity can use at present. Therefore, the problem is not one of a universal energy shortage. It is apparent that the difficulty is rather one of a cultural nature; namely, how do we access this abundant energy resource without producing too much CO_2 in the process?

Among the calls to economize on energy consumption, it is easy to lose sight of the fact that we do have workable solutions available. However, these bring their own challenges for our ways of thinking about and dealing with energy. As an example, I would like to concentrate here on thin-film photovoltaic technology. This technology for the production of available energy has the following radical, new characteristics:

In mid-2009, solar film using photovoltaic technology costs about 50 U.S. cents per 'Wp.' A 'Wp' ('Watt Peak') is a unit used in the measurement of the

#130

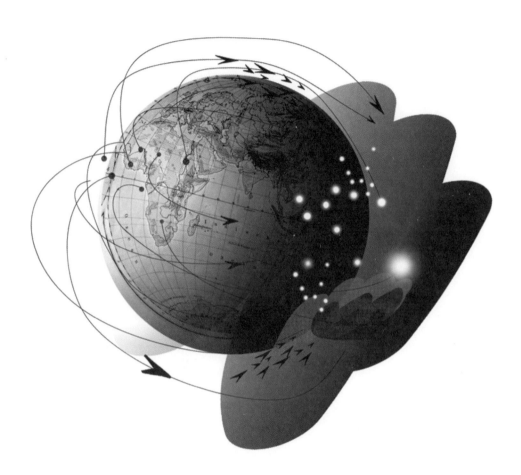

发电机（例如太阳能电池，用黑点标示）将世界各地的电力用户（白点）连接起来，形成具有持续能源的简单系统。

Electricity generators (e.g., solar cells, represented by black points) will be connected worldwide to electricity consumers (white points), forming a single system where the energy is 'always on '

#131

测量能源生产效率的单位。在瑞士，每年每峰瓦可以产生1千瓦时的电。2009年年中，在瑞士每千瓦时电需20美分。归功于新印刷技术的应用，太阳膜的价格每年下降约百分之三十。这意味着到2015年，一个1峰瓦太阳膜将花费不超过0.06美元，然而这些电按现在的标准将创造20美分的价值。因而现今每千瓦时电20美分的市场价在2011年末将会降低。一旦安装了太阳膜，它将提供至少25年干净、清洁的电力。

太阳膜是第一个用于消费者市场的能源生产产品。直到现在，能源生产一直局限于公用事业部门。由于消费者市场的需求强烈，我们期待在能源生产领域有一个更动态的市场。

一个简单的计算表明，一个面积16万平方千米（1/3西班牙）大小的地表就能充分满足整个地球的能源需求。当你把它从整个地球表面积中分离出来时，你就会觉得它所占面积很小。

上述类型太阳能电池板的效率约为10%并可以有所提高。植物绿叶系统吸收太阳能量并将其转换成营养物质的效率为0.2%左右。

我们可以将这种相当粗略的计算制定得更严谨些。很明显我们即将从能源短缺中自我解放出来。太阳能电池是第一个可大规模应用而不需要依靠燃烧产生能量的材料。生产它们需要与生产数据存储设备相同的原料和印刷技术。现在没有人会说数据短缺。我们正在利用因特网

efficiency of energy production. In Switzerland, one Wp can generate 1 kWh of electricity in a year. In mid-2009, 1 kWh costs 20 U.S. cents in Switzerland. The price of solar film is falling by some 30% per year, thanks to the new printing technology being used. This means that by 2015, a 1-Wp solar film will cost less than 0.06 U.S. dollars, whereas the electricity that it will produce—measured by today's standards—will be worth around 20 U.S. cents. Today's market price of 20 U.S. cents per kWh may perhaps be under cut already towards the end of 2011. Once installed, this solar film will deliver at least 25 years' worth of free, clean electricity.

Solar films are the first energy production products to address the consumer market. Until now, energy production has been confined to the public utility sector. Since consumer markets are much more virulent, we can expect a much more dynamical market in the sector of energy production.

A simple calculation shows that a surface area of 160,000 km² (or a third of Spain) would be sufficient to cater for the entire world's energy needs. This is not too much, when you see it divided over the world's surface.

The type of solar panels described above have an efficiency today of around 10% and rising. The system in the leaves of plants that traps the sun's energy and converts it to nutrition has an efficiency of around 0.2%.

The rather cursory calculations presented here can also be formulated more conservatively. It seems certain that we are on the point of emancipating ourselves from energy shortages. Solar cells are the first materials to be available on a large scale that do not have to be burnt to produce energy. And their production needs

解决过剩的数据,比如,用新的文化方式来应对这个问题。能源将在相同的技术基础上工作。基于技术的"千瓦犹如千字节"的呐喊还尚未被完全理解。在相对较短的时间内,我们将拥有过剩的能量,伴随着这种"清洁"能源的发展,我们将开发新形式的文化,但是我们也必须学会避免某些挑剔的使用方式。更加简单地说:能量、能量密度,更重要的是我们使用能源的方式将成为日常生活的附属。我们面临的挑战是生态道德缺失以及如何将生态意识政治化、伦理化。

请允许我以另一种方式来设定这些想法:用一片太阳膜,将它在纽约当地时间14时暴露在太阳下,并在上海凌晨2点时使用虚拟电缆运行。我们将此电缆连接在一个OLED膜上。它将照亮上海,因为纽约此时正在接受日照。再等待12个小时,直到纽约到达夜晚,太阳能电池将不能产生电流,上海的OLED就不再亮了。而上海现在也不需要被照亮,因为那里已经阳光普照了。在我们的虚拟试验中光能可以一直照射在上海,我们再也无需担心没有光照的问题。而我们现在所担心的是如何能够随时将它关闭,比如在我们白天想睡觉时——一个和我们生活习惯完全相反的情形。这就是我们所说的能量生产附属于生活。也就是我们所理解的能量过剩:能源生产几乎是建立在简单的重复生产基础上,而不是建立在有限资源的基础上。

the same raw materials and the same printing technology that we find in our data storage devices. No one today would say that data is, or should be, in short supply. There is a glut of data, and we are developing, with the Internet, for example, new cultural means to cope with this. Energy will come to work on the same technological basis. The slogan 'kilowatts ike kilobytes' has not simply been plucked out of thin air—it is based on a technological reality. In a relatively short time, we will have a surfeit of energy and we will develop new forms of culture along with this 'clean' energy, but we will also have to learn to avoid some particular forms of use. More succinctly put: Energy, energy density, and, above all, the way that we use energy will become mediatized. The challenge will be to leave behind the moralistic ecology of shortages and to develop a political, ethical ecology of surplus.

Allow me to formulate these thoughts in another way: Take a sheet of this solar film and expose it to the sun in New York at 14:00 hours local time, and run an imaginary electric cable to Shanghai, where it is 2:00 am. We connect this cable to an OLED film. It will light up, because the sun is shining in New York. Now wait twelve hours, until night-time in New York; the solar cell is not producing current and the OLED in Shanghai is not lit up. It does not need to be, because the sun is shining there now. However, the light in our imaginary experiment is always on in Shanghai. We no longer need to worry about the light being on. Our concern now is how we turn it off when, for example, we want to go to sleep—a complete reversal of the situation we are accustomed to. This is what we mean when we talk about the mediatization of energy production. This is what we understand by a surplus of energy: energy production almost solely on the basis of reproducible intellect and not on the basis of

当然，位于地球两端的两个元素组成的这个例子并没真正出现在现实中。但是，正如无线电话和互联网一样，由 5 000 亿电力消耗设备和相当数量的发电设备组成一个跨越全球的网络，并不是一个乌托邦式的白日梦。

1969 年，阿波罗 11 号任务让我们第一次从月球上看到了整个地球的全貌。1973 年，黄金本位制度被废除，在同年又爆发了第一次石油危机。今天，我们正将我们的星球置于一个不断精确的传感器和执行器网络（比如谷歌地球）中，正在建设一个在线展现地球地貌的网络代表。我们对地球的态度在很短的时间跨度中发生了改变。一大部分由于气候变化引发的骚乱都归因于我们自己对气候的感受，正如我们当前由于周围环境大规模改变而改变了对地球态度。

这些又和建筑有什么关系呢？Reyner Banham 是第一个将建筑描述为"柔性环境"的人。我们不再围绕着火炉坐在我们的房子里，即使有密闭的窗户和厚实的墙体，也还担心着房间内热量会流失。能量变得越来越容易掌控。我们居住于一个气候条件网络中。

要使人真正理解这一点：在苏黎世 19 世纪房子里增加隔热层和在西班牙装太阳能跟踪系统哪个更便宜？即使从中期看来，也绝对是后者！

limited resources. Of course, the example with the two elements on opposite sides of the world does not really work. But a globe-spanning network composed of 500 billion electricity-consuming devices and a similar number of electricity generators is no utopian pipedream, as mobile phones and the Internet have already shown.

In 1969, the Apollo 11 mission allowed us, for the first time, to look back from the Moon and see the Earth in its entirety. The gold standard was abolished in 1973, and in that same year the first oil crisis took hold. Today, we have wreathed our planet in an ever-tightening net of sensors and actuators and are constructing—with Google Earth, for example—an online representation of our planet from the inside out. Attitudes toward our planet have changed within a very short time span. A large part of the present commotion about climate change must be attributed to the proprioception-like attitude we currently develop towards our planet, induced by the large-scale mediatization of our surroundings.

So what does all that have to do with architecture? Reyner Banham first described architecture as a 'well-tempered environment.' No, we no longer sit in our houses around fires, with our airtight windows and thick walls, worrying about the heat escaping. Energy is becoming increasingly mediatised. We inhabit a network of climatic conditions.

To really drive the point home: Which is cheaper, the additional insulation in a 19th century house in Zurich, or a solar tracking system in Spain? Even in the medium term, the latter, surely!

时　间： 2006—2009
参与者： Carsten Droste
合作者： Schweizerische Zentralstelle für Baurationalisierung CRB (CH); Byron Informatik AG (Basel, CH); futureLAB AG (Winterthur, CH)

瑞士一切网络化 All of Switzerland Online

"CRB 在线"项目是除了 digitalSTROM 外我们部门第二成功的大型基础设施项目。该项目使用有高精度细节的 web2.0 系统，拥有 50 亿瑞士法郎的投资，通过该项目，我们已经可以成功地代表包括市场容量的所有瑞士建设活动。现今世界范围内还不存在类似系统，我们为能在 2009 年 5 月将它上线而感到自豪。

这个为未来建设提供了一个完美的平台的项目是怎样的呢？这个项目的设计师是 Schweizerische Zentralstelle für Baurationalisierung，CRB（瑞士合理化建设中心），是一个由建筑师、工程师机构、建筑工人、建筑公司、结构和土木工程部门组成的团体，几乎囊括了整个瑞士的建筑业。CRB 的目标是建立和运行建筑师和工程师对建筑设计品质精确设想的基础性设施，使得分包商和企业知道需要做什么，提出相应报价并且顺利执行合同。为此，在其存在的 50 年里，CRB 用三种瑞士民族语言——德语、法语、意大利语制定了超过百万件所谓的交付方案。由此所产生的标准目录已经相当成功，并被 80% 的瑞士建设事业所采纳。

Next to digitalSTROM, the 'CRB online' project is the second successful large-infrastructure project from our department. With this project we were able to successfully represent all the building activities in Switzerland—comprising a market volume, after all, of 50 billion Swiss francs—with a high degree of detail in a web 2.0 system. No other comparable system exists today, worldwide, and we are proud of having taken it online in May 2009.

So what is this project—which presented a perfect platform for many further developments—actually about? The project owner is the Schweizerische Zentralstelle für Baurationalisierung CRB (the Swiss Center for the Rationalization of Construction CRB), an umbrella organization comprising the institutes for architects, engineers, construction workers, construction firms, and the structural and civil engineering sectors, which represent almost the entire Swiss building industry. The goal of the CRB is to build up and run an infrastructure that will describe precisely the qualities envisaged for the buildings created by engineers and architects, so that subcontractors and firms can understand what needs to be done, to structure their quotes accordingly and to run their contracts successfully. To this end, in the 50 years of its existence, the CRB has formulated over one million so-called 'deliverables' in three of the national languages of Switzerland, namely German, French, and Italian. The resulting Normpositionenkatalog (NPK, Catalog of Standards) has been quite successful, having been consulted in around 80% of the building work in Switzerland.

但是，现在这个系统的发展面临许多压力。我们很容易想象从所提供的众多目标中选择合适的"交付方案"是多么困难的一件事。尽管有许多描述说明，也时常会找不到适合的"交付方案"，何况它们的分类也不够明确。更关键的是，一个建筑项目的性能规范在规划阶段必须被描述得非常详细，也就是说，费用必须在规划阶段就确定下来。此外，这种高度详细的成本规划增加了政府的负担。现在缺少的是与规划阶段接轨的循序渐进的成本估算体系。这将从一开始就将规划和预算体系不失灵活性地紧密连接起来。现在，在数量和成本的重要性超过了质量的形势下，这将成为主要目标。

对于这样的任务类型，一些集团和各级组织允许交付说明被分解为多个阶段。建筑由许多组件组成，比如外墙窗户由框架、安装、清洁和维护依次构成。通过使用这些组件，甚至是在早期的规划阶段，都可以快速估算出成本。例如，一个典型的砖墙外立面幼儿园就可以作为一个样本，而不需要再去进行新的混凝土表皮设计。

我们可以创建一个具有所有可能性的建筑规划书，在瑞士，这种规划大约有100万个条目长度。但是问题恰恰也是出现在这里，不是所有建筑的轮廓都能被建造出来。各个行业、建筑项目、开发商的术语都大不相同。即使是在瑞士这样一个小国家，各地的

However, this system is now finding itself under pressure. It is easy to imagine how difficult it is to choose the right 'deliverable' for a building project from the many on offer. In spite of the large number of specifications, it is often not possible to find a suitable one, and sometimes their classification is not sufficiently explicit. More importantly, however, the performance specification of a building project can only be described at a very detailed level after the planning stage; i.e., the costs can be determined only when the planning is concluded. Additionally, with its high level of detail, the cost determination creates an enormous bureaucratic overhead. What is missing is a stepwise cost estimation to go hand-in-hand with the planning stages. This would closely link the planning and cost estimation from the outset, without losing flexibility. Nowadays, where quantities and costs are overtaking quality in importance, this should be a central goal.

For this type of task, certain groupings and levels of organization allow the delivery specifications to be broken down in stages. A building consists of—among other things—a facade, which is composed of windows, in turn comprising the frames, installation, and cleaning and maintenance. By using these groupings, quick cost estimations can be made, even in the early planning stages. For example, the cost of a typical brick facade for a kindergarten can be used as a rule of thumb, without the specific details of the new concrete facade having been planned.

A list of all possible specifications for a building can be created; as described above, in Switzerland, this list is some one million entries long. But—and herein lies the problem—an outline across all buildings cannot be constructed. The pragmatics of the various trades, building projects, and developers are simply too different. Even in a small country like Switzerland, building cultures

All of Switzerland Online

苏黎世能源的监测：红色标识表示高能源使用，绿色标识表示低能源使用。

每个建筑都包含自己的相关施工信息，可以在设置相关访问权限的基础上将其用于结构评价，或是作为新建筑、翻修的经验基础。

The city of Zurich as an energy monitor: red signals high energy use, green signals low energy use.

Every building 'contains' its own relevant construction information and can be used—with relevant access permissions—for structural evaluations or as an 'experience base' for new builds or renovations.

建筑文化也都大不相同，并且建筑的语义和建筑技术发展得过快。一个单一的分类体系将严格管制建筑发展进程，将其确实锁定在一个可预见的计划中。在过去，我们一直在尝试建立这样一个标准化系统，但收效甚微。当前一个突出的例子是工业基础类（IFC）。因为使用困难，开发商通过补充更多细节继续将其完善，但却使得其更加难以利用。最终导致了这个进退两难的局面。

这就是我们的 CRB 在线系统的设计来由。该系统的一致性不再是通过制约用户使用同一种分类方式来获得。在 CRB 在线系统中，我们提取了分类。现在最终用户可以为他们个人建筑项目生成各自的订制分类。由于它不是对所有建筑类型负责，因而现在每种分类都快速、紧凑、实用。CRB 在线系统网络平台是第一个在不同使用者、项目、地区、时间段的多种分类中建立交流的系统类型。用户们无需在事先决定采用相同分类的情况下就可以交互式地分享他们的经验。个人自由和群体共识之间传统的目标冲突在一个新的、抽象的领域被解决了。CRB 在线系统提供了一个做梦也想不到的可能，让建筑师和他们的合作伙伴在互联网进行合作，因而没有传统系统强加于个人自由的制约。

该系统于 2009 年 5 月上线，并用市场上最流行的软件提供更新。针对传统保守的建筑行业的

are by far too different. and, over the years, the semantics of building, as well as the technological underpinnings, develop too quickly for a printed catalog. A single classification scheme would regiment the construction process, actually locking it down for the foreseeable future. In the past, there have always been attempts to establish such a stadardized system—yet with little success. A current salient example is the International Foundation Classes (IFC). Because they are so difficult to use, their developers continue to refine them by adding ever more details, in turn rendering them even more difficult to use. A no-win situation.

This is where our CRB online system comes in. The consistency of the system is no longer achieved by constraining the end user to a one-size-fits-all classification. With CRB online, we are abstracting the classifications. Now, end users can generate their own customized classifications for their individual building projects. Since it does not have to do duty for every building type, each classification is now fast, compact, and practical. The CRB online Internet platform is the first system of its type to establish proper communication between various classifications of different users, projects, regions, and time periods. Users can interactively share their experiences without having to decide upon a common classification beforehand. The traditional conflict of goals between individual freedom and communal consensus is resolved in a new, abstract sphere. CRB online releases an undreamed-of potential for Internet-based co-operation between architects and their partners, without the constraints on individual freedom imposed by conventional systems.

The system went online in May 2009, and was delivered with updates for the most popular software packages in this market. For clients in the

客户，也尽量减少变化。分类和交付说明目录通过在线连接链接到 CRB 服务器。新系统模仿旧系统，采用使用者的旧习惯推出新的发展以作为媒介。从现在开始，加快工作和新商业模式的发展速度不仅仅是使用者的需求也是市场动态的需求。CRB 可以调节这一进程。

traditionally conservative construction industry, the changes have been minimal. Classifications and delivery specification catalogs are accessed via an online connection to CRB's server. In its appearance, the new system mimics the old, with new developments using the user's old habits as their medium. From now on, the speed of developing new methods of working and new business models is simply a function of the desires of the user and the dynamics of the market. CRB will only moderate this process.

我们可以想象许多可能的情景，比如：能源运营数字的计算与规划成本估算同时进行，拓扑与几何描述共同运作，或者同时反映出数量和质量。我们可以在规划和建造任务中"搜索"，寻找类似相同项目。现在的情况是，我们再也不需要查阅案例研究或是书籍。工作将变得更加不同。建筑不再只单独通过能源相关的问题进行比较。规划将联系成网络。项目之间的壁垒将变得模糊。

We can imagine many possible scenarios; for example: energy operating figures being calculated alongside planning cost estimates, or topological and geometrical descriptions being worked on and indexed alongside the qualities and quantities. We could conceivably 'google' the state of building in Switzerland and, during planning or construction tasks, look for projects that are similar to ours. One would no longer need to consult case studies or books, as is the case today. Job descriptions would change and become more differentiated. Buildings would no longer be explicitly compared according to energy-related questions alone. Planning would be networked. The barriers between projects would blur.

2000 年，我们以"回到现实"的箴言作为开始，因为我们都不想耗费大量精力在数字建筑模型和它们富有诱惑力的三维表现上。具有讽刺意义的是，我们现在正在使用虚拟现实技术来尝试持续代表瑞士全部建筑业资本，以便于将所有与规划和建设过程相关的事项都放在合适的位置，以满足他们各自所需的特殊背景。

In the year 2000, we began with the motto 'back to reality,' since we did not want to expend all of our energy on digital architectural models and their admittedly seductive three-dimensional representations. Ironically, we now find ourselves in the position of trying to consistently represent the entire building stock of Switzerland using virtual reality, in order to put all those involved in planning and construction processes into the position to actualize what they need according to their specific and singular contexts.

第八章

虚拟现实的应用

　　这本书是我在 CAAD 实验室在职期间前三分之一时期工作的概述。在这八年间我们进行了 100 多项实验，我们逐渐树立了信心，已经大概了解了信息技术在建筑学领域内的应用可能——至少在那些现在已有理论支持的范围。在这些领域，我们特别进行了深入和广泛的工作。在下一步中如何将我们的工作成果优化、定位，使之能够应用于实际工程，就需要相关的经济学专家和市场销售专家来配合了。各种相关的显示证明这一步是非常有意义的。现在，我们就要提出这个挑战性的问题：我们 CAAD 小组下一个阶段的工作是什么？

　　当然，我们会继续进行实验。与艺术史或建筑施工相反，我们的研究领域发展变化得很快。当我们的研究在 2000 年底开始时，谷歌建立还不到两年，维基百科成立也只是一年之遥。这就是为什么我坚持这样一个实验性、跨学科的小组，无需遵守什么理念，没有什么等级制度，有的只是一颗好奇心，能够敏锐地感知新的发展，提出恰当的问题，给出有趣的答案。有力的领导和有限的假设考虑到产生结果，也许在短期内会很有效。但就中期和长期来看，这个方法必定失败，因为它缺少促成发展的对深层机构的判断和认识。

　　我们还会在现有研究成果上继续探索。我们创办的公司现在已有 70 个 ETH 以外的人加入进来。可喜的是，公司发展得越来越强大。新技术的开发和商业运作模式都显示了巨大的潜力！

　　为了继续保持发散性实验的方式，从已有实验中总结将是至关重要的，并进一步研究方法和理论依据，以使我们能够研究探索这些应用在实际中会发挥多大的功效。但是如何才能实现这一步？要构想出一个什么样的理论，才能使当代信息和网络技术被认可并应用，尤其是对比另外一些我们知道的破坏性的技术更快地应用？大众对于技术或媒体理论的讨论，总是以一种回顾反思的方式，而且，由于这些领域的发展速度过快，理解和参与到当前的发展趋势中都非常困难。例如，由有机材料印刷所引发的电子生产技术的巨大跨越，很少或基本没有相关的理论讨论。但是，如果我们仅仅从这些技术所带来的量化的结果来看——以能源生产为例（见"能源过剩"一节）——我们可以模糊地看到这种新应用所带来的强大冲击力，在非常智能的一些基础设施中，在其他一些领域都可以看到。今天，在大多数情况下，基础设施的组织还遵循着机械的管理体系。建立供水的大型水库属于这一类，就像许多其他的例子，比如生产电力的核电站、运输燃料的欧洲大陆石油管道、闭路视频安全监控网络、建立在正交网格上的旨在缓解住房短缺的巨大的新城市，以及旨在增产的大规模集约耕种的土地。

　　这种有关基础建设的思维方式是建立在保护自己以应对大自然的变化以及能源短缺的基础上。这种随着技术控制系统而进行基础设施优化的行为是符合工业时代特点的；它还有一个前提，那就是认为用信息技术为传统基础设施进行的改良大部分已经实现。通过不断地优化和交联，另一种情况也在同时发生，那就是原本被认为是提高了效率的活动从一个非常长远的角度来看变得非常脆弱。这些最初作为对抗恶劣自然环境并起到保护作用的构想，需要更加复杂的控制水平来操作，这使它们相应的技术变得或复杂得离谱，或专制异常，或两者皆有。

我们的做法是看信息技术有多大的能力来更新我们的基础设施，而不是在它原有基础上继续优化它。相反的，我们希望这项技术有一种新的效益，而且在概念上具有灵活性；在今天，概念和标号总是走在价值和收入之前。这种灵活性让我们能够从根本的角度反思传统的解决问题方式，传统方式最初是用来保证基础设施正常运作的。但如果我们坚持把新型的基础设施仅仅看成是在和大自然作斗争时提供必需品的装置的话，它们将不能被充分理解和定义。相反的，我们希望有一种完全不同于"必须"的概念，这会给我们在一个全新的层次上开展一个全新的游戏。新型的基础设施是一种新的概念系统，在这里，每一个问题都有多种的解决方案。从一个更积极的角度来看，它们——保有原有的供应功能——成为能动者，是新的行为习惯的创造者。相比于目前解决问题的程式化来说，它为我们提供了新的自由，基于数据层级的自由。我们谈论的是"记叙性"的基础设施。在这一建筑学新领域中，最明显的特征就是"没有标准的建筑"。技术，被认为是发现这一自由的眼镜，却并没有被充分利用。我们认为，这就是技术、经济、文化最具开发潜力的地方。

这片新的领域是随着信息和通讯技术成为基础设施而出现的。它改变了我们处理问题的方式，传统的方式遵循着盛行的、面对物理环境"将有限资源最大化利用"的原则。相比之下，新的基础设施需要我们遵从的是处理丰富的原则——数据的丰富。在 2002 年，人们消耗了 2EB 的数据。到 2009 年，这一数字已经变成了惊人的 250EB。此外，在 2009 年，产生的新数据比人类历史上所公布的还要多。所有这些数据都有潜在的意义和作用，正因如此，显示了一种不断增多的资源，就如同实证研究的一种准自然。最终，我们知道了为什么信息网络成了新行为、新习惯、新工程出现和发展的平台。谷歌、维基百科、脸书、推特、短信和全球定位系统都可以作为一个典型的例子。它们在基础建造行业与新的信息技术行业之间建立一个很好的交流平台（见"瑞士一切在线"一节）。

到现在为止，多亏了我们对于虚拟现实可以有无穷多的概念性的审视态度，和令人惊讶的纯数据处理的能力，我们的工作才可以打着"回到现实"的旗号，达到把它植入现实生活的目的。虽然从一开始，我们就很清楚"让建筑离开电脑"这条格言只会存在很短的时间，但我们仍有意识地做出决定，不开发一套方法使我们只用轻轻点一下鼠标，就生成整个瑞士的规划。同样，对于场地分析，我们没有预先打包数据，我们没有植入控制系统使自己能够控制瑞士的活动，并且我们并没有试图教育瑞士公民处理个人的能源需求。我们的搭建环境中，有很多这样便捷的有上述功效的应用程序。但我们有意地把我们的项目定位于不同的功能。我们使用相同的技术，在实际的地点给实际的工程注入更大潜力，这样它们就能以一种完全不同于之前的形式呈现。仅此而已。

这也将对我们的理论建设有个新的定位。这个理论，不加以解释，它不能为自己解释，而我们要在更好地理解、更大的信心的基础上来处理这更大的自由。显而易见的是，这个理论会与现实生活中逐渐增多的附属品有很大联系。从我们的角度来看，这与增加世界的可控性、人工性、虚拟性并无多大关系，而更多与日常生活的变化息息相关，它通过新的灵活度、速度和社交媒介所产生的各种接触渠道来展现自身。

换句话说，这种发展可以理解为我们行为深度的延伸：如同从原子到分子，从几何方法和机械操作技术到统计方法和工程建模。对于基于信息技术和网络的新型基础设施，我们需要新的理论观点来帮助我们培养运用各种具有潜力的工具，在允许的情况下，来合理地解决各种不同情况。

　　为此，我们最近成立了虚拟现实应用研究所。这种大胆的抽象分离——与特定的"信息技术和搭建环境"的背景分离——刺激我们提出了重要的主题，比如能源、环境、金融和教育。不像我们以前的实验，总是被定位于基于现实的当地建筑实践的研究，我们现在想进行一种国际性以及跨学科的课题，讨论我们自己提出的问题。借助我们的背景，我们可以提出一个明确的观点：我们想要开启一种全新的共同渠道，并在全世界寻找能够倾听我们的观众。

　　这就是我这八年在苏黎世联邦理工大学当教授所积累下来的经验。我希望我们已经激起了你的兴趣。我们期待着下一个八年，在我们面前展开全新的、开放的研究领域！

Applied Virtuality

This book provides an overview of our work at the Chair for CAAD after the first third of my period. With more than 100 experiments over a period of eight years, we are gradually becoming confident of having gained a certain overview of the possibilities for utilizing information technology in architecture—at least amongst those theoretical perspectives currently available. We have investigated and worked extensively on those realms. The further optimization and positioning of our outcomes as real-world applications is now a matter for professionals in economics and marketing. The various spin-offs show where such a step makes sense. We now have to pose the challenging question: How should our work at the Chair for CAAD during the next period of eight years be oriented?

Certainly, we will continue conducting experiments. In contrast to art history or building construction, our field of study is developing and changing rapidly. When we started at the end of 2000, Google was not yet two years old, and the founding of Wikipedia was a year away. This is why I will retain an experimental, interdisciplinary team with limited preconceptions or hierarchy, driven freely by a curiosity best suited for identifying new developments, asking the right questions, and coming up with interesting solutions. Hierarchies and narrow presumptions may be more productive in the short term, as far as results are concerned. But in the mid- and long-term, this approach would fail because the integral developments at a deep structure level could not be diagnosed and recognized.

We will also continue to endeavor to spin off start-up ventures from the results of our research. The companies we have started so far employ around 70 people outside of the ETH Zurich. Hopefully, these companies will go from strength to strength. The technologies developed and the business models driving them show much promise!

In order to keep on executing experiments in a radical way, it will be vital to abstract from those carried out until now, and to develop methods and theories that will allow us to investigate and explore the efficacy and power of these possible applications in real contexts. But how to achieve such a step? How would a theory have to be conceived in order to make the new potentials of contemporary information and network technologies available and, importantly, more quickly available than what we know from other cases of disruptive technologies? The general discourse in the theory of technology as well as in media theory, is often orientated retrospectively, and, due to the speed of development in these areas, seems to have great difficulties in understanding or even anticipating current tendencies. It is, for example, striking that the leap in the production technology of electronics brought about by printing with organic materials seems seldom, if at all, to be discussed in the relevant theoretical discourses. But if we think of the purely quantitative consequences of these developments—in the case of energy production, for example (Energy Abundance → pp. 236 ff.)—we get a vague idea of how drastic the effects of this new availability, in the sign of smart infrastructures, will be also in other areas. Today, the organization of infrastructures still follows, in most cases, the scheme of mechanically controllable hierarchies. The massive dams erected for water supply belong in this category, just as, among many other examples, nuclear power stations for the production of electricity, transcontinental oil pipelines for fuel delivery, closed-circuit video surveillance networks for security, huge new cities built on grid systems and designed to relieve housing shortages, and massive intensively worked fields intended to increase food production.

This way of thinking about infrastructures is based upon the idea of safeguarding ourselves against primary shortages and the vagaries of nature. The so conceived infrastructures were optimized as technical control systems according to the perspective of the industrial age; it is also within this mindset that most of the upgrades based on information technology for these traditional infrastructures have been realized. Through refinement and cross-linking, this has meanwhile taken place to such an extent that their initially increased efficacy has, if considered within a fairly long-term perspective, become surprisingly fragile. These systems, originally conceived as a protection against the inhospitable face of nature, need ever more complicated levels of control that render their corresponding technologies either monstrous or tyrannical—or both at once.

Our approach is to see the potential of the information technology upgrades to our infrastructure not in terms of optimizations to the functions they already possess. Rather, we expect a new efficacy for this technology growing out of the flexibility within the symbolic; today, streams of signs and symbols are running on ahead of streams of values and revenues. This flexibility allows for a more fundamental way of rethinking the traditional problem-solving approaches that used to guarantee the functioning of infrastructure. These new infrastructures would not be adequately characterized if we continued to insist on seeing them as nothing more than suppliers of 'the necessary' in our struggle with nature. Rather, there will be a differentiation of 'the necessary' that will open up a new game at a whole new level. These new infrastructures will become symbolic systems in which for every problem several possible solutions can be found. In an ever more active sense, they will become—alongside their function as supply systems—'enablers' for new habits. They reflect the new freedoms that, relative to the hitherto existing formulations of problems, exist on a meta-level. We are talking about 'narrative' infrastructures. The most obvious architectural incarnation on this new plateau is the so-called 'Non-standard Architectures.' Technology, understood as a looking glass onto this new freedom, feels completely unfamiliar. In our opinion, this is where the technological, economic and cultural potentials are to be found.

This new plateau emerges as information and communications technologies become infrastructure. It changes the traditional focus of technology according to the principle of 'optimized handling of limited resources' that had prevailed vis-à-vis a physical nature. By contrast, the new infrastructures confront us with the principle of a dazzling abundance—an abundance of data. In 2002, humankind had at its disposal over two Exabytes of explicit data. By 2009, this had become a staggering 250 Exabytes. Moreover, in the year 2009 to date, more data has been generated than has been spoken in all of human history (60 Exabytes). All this data is potentially meaningful and useful and, as such, manifests a continually growing resource that can be seen as a sort of quasi-nature for empirical research. It is becoming clear, at last, why information networks are providing a platform for the development of new behaviors, habits, and projects. Google, Wikipedia, Facebook, Twitter, SMS, and GPS can serve here as an example, as can the translation of the principles of these new information-handling infrastructures to the construction industry (All of Switzerland Online → pp. 242 ff.).

Until now, thanks to our skepticism of the bottomless symbolism of virtuality and out of respect for the incredible performance of pure data processing, our work had gone under the banner of 'Back to Reality,' with the intent to anchor it within the concreteness of real-life situations. Although it was clear from the outset that the maxim 'Get Architecture Out of the Computer!' would only be useful for a limited time, it was a conscious decision not to develop a grammar that would allow us, at the flick of a switch, to plan and develop the whole of Switzerland. Similarly, we had compiled no statistics for site analysis; we had conceived no control system with which we could monitor movements within Switzerland; and we had not sought to educate Swiss citizens on their individual energy needs. There are many of these quick and allegedly effective applications of information technology in our built environment. We have purposely formulated our projects differently. We have used the same technologies to infuse concrete projects in concrete locations with more potential, so they can unfold and develop in a more differentiated manner than they would have previously. No more, and no less.

This will also be our position for the new vector of theory building. A theory that allows us to deal with these freedoms on the basis of a better understanding, and with more confidence, must be orientated by something it cannot itself explain. Obviously, this theory will concern itself with the virtual and with the increasing mediatization of our environment. From our point of view, this has less to do with reduplication in controllable, artificial, illusory worlds, and more with those changes in every day life that announce themselves through the new flexibility, speed, and access options created by the communication media.

Put another way, these developments could be described as an extension of our scope of action: from the atomic to the molecular level, from geometric methods and mechanically operating technology to statistical methods and a technology of projective modeling. For the new infrastructures based on information technology and networks, new theoretical concepts are needed to help us cultivate the power of technologies that are effective within the realm of the potential, as well as to cope reasonably with the differentiations these technologies allow.

To this end, we have recently founded the Institute for Applied Virtuality. This step into abstraction—departing from our specific background in 'information technology and built environment'—dares us to broach such important themes as energy, environment, finance, and education. Unlike in our previous experiments, where it was important to anchor our research within an architectural practice that is always concrete and local, we now want to engage in an international, as well as an interdisciplinary discourse on the questions we have formulated. With our backgrounds, we can formulate a specific viewpoint: We want to confidently open up new channels of communication and find an international audience for our ideas.

Well, those were the first eight years of my professorship at the ETH in Zurich. I hope that we have piqued your interest. We are looking forward to the next eight years and to the new, open fields of investigation spreading out before us!

Isabelle Stengers:
The Invention
of Modern Science

' Computer simulations not only propose an advent
of the fictional use of mathematics,
they subvert equally the hierarchy between the purified phenomenon,
responding to the ideal intelligibility invented by the experimental
representation, and anecdotal complications. '

Staff Members 2000–2009

Ulrike Bahr
Mathias Peter Bernhard
Dr. Li Biao
Katharina Bosch
Markus Braach
Dr. phil. des. Vera Bühlmann
Rodrigo Omar Carrizo Couto
Benjamin Dillenburger
Philipp Dohmen
Maja Dzieglewska
Karsten Droste
Cornelia Erdmann
Neil Franklin
Pia Fricker
Oliver Fritz
Dr. Andrea Gleiniger
Mario Guala
Michael Hansmeyer
Dr. Mikako Harada
Marcus Hsu
Dr. jur. Martin Jann
Alexandre Kapellos
Christin Kempf
Dr. Toni Kotnik
Dr. med. Patrik Künzler
Silke Lang
Lee Seong Lee
Alexander Lehnerer
Steffen Lemmerzahl
Russell Loveridge
Jonas Mahrer
Sebastian Michael
Georg Munkel
Dr. Christian Müller
Mathias Ochsendorf
Maria Papanikolaou
Tom Pawlofsky
Jörg Pottrick
Bryan Raney
Denis Raschpichler
Werner Riniker
Aoife Rosenmeyer
Saikat Roy
Kai Rüdenauer
Philipp Schaerer
Fabian Scheurer
Christoph Schindler
Odilo Schoch
Susanne Schumacher
Thomas Seibert
David Sekanina
Torsten Spindler
Sibylla Spycher
Kai Strehlke
Andrew Vande Moere
Dr. des. Georg Vrachliotis
Dr. Steffen P. Walz
Christoph Wartmann
Klaus Wassermann
Cecile Weibel
Oskar Zieta

图书在版编目（CIP）数据

超越网格：建筑和信息技术、建筑学数字化应用／（德）豪威斯塔德（Hovestadt, L.）编著；李飚等译．—南京：东南大学出版社，2015.10
（建筑设计数字技术研究丛书／李飚主编）
ISBN 978-7-5641-5709-8

Ⅰ．①超… Ⅱ．①豪… ②李… Ⅲ．①数字技术—应用—建筑学—研究 Ⅳ．①TU-0

中国版本图书馆CIP数据核字（2015）第091190号

书　　名：	超越网格——建筑和信息技术、建筑学数字化应用
著　　者：	Ludger Hovestadt
译　　者：	李飚　华好　乔传斌
责任编辑：	戴丽　魏晓平
责任印制：	张文礼
出版发行：	东南大学出版社
社　　址：	南京市四牌楼2号　邮编：210096
出 版 人：	江建中
网　　址：	http://www.seupress.com
印　　刷：	利丰雅高印刷（深圳）有限公司
开　　本：	700mm×1000mm　1/16　印张：17　字数：408千字
版　　次：	2015年10月第1版
印　　次：	2015年10月第1次印刷
书　　号：	ISBN 978-7-5641-5709-8
定　　价：	146.00元
经　　销：	全国各地新华书店
发行热线：	025-83791830

本社图书若有印装质量问题，请直接与营销部联系。电话（传真）：025-83791830